D1085003

GRASS FARMING

GRASS FARMING

by

M. McG. COOPER

C.B.E., B.Agr.Sc.(N.Z.), Dip.Rur.Econ. (Oxon), B. Litt. (Oxon), F.R.S.E.

Formerly Dean of Agriculture, University of Durham, King's College, Newcastle-upon-Tyne

and

DAVID W. MORRIS

B.Sc. (Wales), Ph.D. (Dunelm), F.R.Ag.S.

Principal, Welsh Agricultural College, Aberystwyth

FARMING PRESS LTD.
WHARFEDALE ROAD, IPSWICH, SUFFOLK

First published 1961
Second impression 1962
Second Edition (Revised) 1967
Third Edition (Revised) 1973
Fourth Edition (Revised) 1977
Second impression 1979

ISBN 0 85236 076 2

This book is set in 11 pt. on 13 pt. Times Roman and printed in Great Britain on Longbow paper by Page Bros (Norwich) Ltd, Norwich

CONTENTS

DIAGRAMS

PLATES

FOREWORD

BY A. S. CRAY M.B.E., J.P.

SINCE Professor Cooper gave us *Grass Farming* in 1961, the farming scene has undergone important changes. In their Preface, the authors of this revised edition have called attention to several of them.

They mention the great advances in technology and in the management of pastures and grazing animals, and there is of course significant improvement in the returns from beef and sheep as compared to milk. Ability to exploit the grass crop is vital to enable us to be competitors in the Common Market, to fill the gap caused by reducing imports of meat and animal products and to meet the needs of an increasing population.

It is still the position that efficient grass production and utilisation is the most complex of our agricultural activities. The comparison between the tillage crop and the grass crop which was made in the Foreword to the 1961 edition still holds good. With the tillage crop the management of the single plant is simple compared with the grass crop, where is it usual to grow two or more varieties together. Each has a different growth habit and each is responsive to or is depressed by certain treatments at certain times. A treatment beneficial to one may depress the other. The crop is growing and is being harvested intermittently or continuously over a long period, with consequent difficulty of accurate assessment of annual yield. Dealing with these challenges is what grassland management is all about.

In this edition the authors have dealt with virtually all the circumstances that arise in the day-to-day and year-to-year management of grassland of all types. Even more important perhaps, they have covered these features of animal management that are related to the successful conversion of grass, fresh and conserved, to meat and milk. Failure in this final stage means failure of the whole enterprise.

It has long been apparent that the loss of nutrients in conservation as hay or silage is the outstanding weakness of our grassland husbandry. The need is for a method of conservation that approaches the efficiency of grass drying but at lower capital cost. The authors call attention to the heavy nutrient losses involved in the making of silage and hay but they have given detailed guidance on keeping these losses to a minimum.

Farmers will appreciate particularly the farming language used throughout this book and the clear-cut separation of the various aspects of their subject. This, and the clear indexing, add greatly to its value as a practical handbook to be referred to time and again after the first reading.

Professor Cooper and Dr Morris are to be congratulated on their success in bringing up to date a book on this all-important subject. It will be welcomed and appreciated by an ever-growing number of farmers and farm managers who have come to realise the potential of the grass crop.

A. S. CRAY

Roe Downs Farm,
Medstead,
Nr. Alton,
Hants.
May, 1973

PREFACE

WHEN the senior author prepared the manuscript for the first edition in 1961, entry into the Common Market was no more than a distant and uncertain prospect. Now it is a reality and British agriculture is facing a challenge at least as great as that created by the Second World War. But in 1939 British farming was in poor shape. Millions of acres had either gone out of effective production or were in low quality permanent pasture, but today the lands of Britain are in good heart and there is a level of technology that compares favourably with that of any country in the world.

In particular, there has been a tremendous advance in the management of pasture and in the handling of grazing animals. Though agricultural scientists have contributed to these advances, unquestionably the major contribution to the rise in the general level of grassland husbandry in Britain is attributable to progressive farmers that can be found in almost every county in this country. We have in mind men like Sam Cray in Hampshire, Ted Owens in Somerset, Edwin Bushby in Cumberland, Oliver Barraclough in Yorkshire (on difficult millstone grit that would break most people's hearts) and Maitland Mackie Junior in Aberdeen. These are but representative names of the many who have convinced their fellow farmers that pasture is green gold if it is properly handled. It is to men like these that this book is dedicated and also to the memory of that very great agricultural scientist, Sir George Stapledon, who back in the thirties realised before any of his contemporaries the importance of good grassland husbandry in the entity of British farming.

We would like to thank everyone who has helped us in the preparation of this book. We cannot mention each one by name but we would be most ungrateful if we did not say 'thank you' to Senorita Teresa Caro Santa Cruz and Mrs Beryl Morgan for deciphering our untidy long hand to put it into shape for the typesetters. Our long suffering spouses—the wife and daughter of the senior author—also deserve a mention because authorship imposes such restraints on family life.

M. McG. Cooper
Holme House
Lesbury
Alnwick
Northumberland

D. W. Morris
Felin Gyffin
Dole
Bow Street
Aberystwyth

CHAPTER I

THE IMPORTANCE OF PASTURE

PASTURE is the most important single crop in British agriculture. In its various forms, short and long leys and permanent grass, it accounts for nearly two-thirds of the cultivable land and, in addition, there are approximately 6.8 million hectares of rough grazings which make an important contribution to output in an area where there is a critical shortage—namely, the production of meat. When the history of British agriculture is written to cover that dynamic period which started in 1939, and which one hopes will not come to an end, the contribution of these so-called rough grazings will be really appreciated. In recent years there has been criticism that too much money, mainly in the form of Government subsidies, has been put into the upland sector where most of the rough grazings are to be found, but lowland farming in Britain, and the British housewife too, would be in a much worse position were it not for the contribution that these marginal areas make to the nation's food supplies.

If there has to be criticism about upland farming, in our opinion it would be better directed to the failure to put the necessary research effort into the betterment of these lands to achieve a fuller realisation of their production potential. Unquestionably they have a very much greater contribution to make in terms of feeding cattle and store sheep now that we are facing the reality of life in the Common Market.

Foresters may say that this land has another function but no one has yet looked critically at the economics of afforestation and compared them with the more immediate results that can be obtained by a sensible input of capital and technical know-how into the improvement of this land for livestock production. The recent rise in the value of lowland farms puts our uplands into a totally new perspective and though it can never be land that will produce high priced crops it can make a very vital contribution to meat supplies, now that a sirloin steak is as expensive as fresh salmon.

LOWLAND GRASS

We must not, at this early stage, let our enthusiasm for the uplands and their developmental problems obscure the importance of lowland grass—the 2.3 million hectares of rotational grass and the 5 million hectares of permanent grass that account for such a significant proportion of total agricultural production. There was a time, and we will be referring to this topic in greater detail later, when permanent grass was regarded as a term of abuse, and great endeavours, mainly in the form of generous ploughing-out grants, were made to reduce its extent. These grants substantially failed, not because the carrot was insufficiently attractive, but because of a general realisation that permanent grass, when it is properly managed, can be just as productive as leys and is generally a cheaper source of nutrients than temporary grass. Particularly is this true on heavy land where the soil can lose structure when it is under the plough. Under these conditions, too, permanent grass is especially valuable, in comparison with short-term leys, when there is high intensity stocking because damage from poaching can rapidly become a limiting factor to productivity.

Apart from their contribution to the extension of the grazing season, the greatest contribution that leys make in British agriculture lies in the effect they have on yields of arable crops under a system of alternate husbandry. There is a very limited range of soils in Britain, mainly brick earths, and silty loams, where continuous arable cropping can be maintained without serious detriment to yields. A few years ago, in a vain endeavour to simplofy farming and to reach the happy situation of a five-day week, except at times of pressure like seeding and harvest, a large number of farmers turned to continuous cereal growing on land that was manifestly unsuited for such management.

Unfortunately they had the encouragement of advisory services which had been bitten by the bug of gross margins and capital was re-deployed from an investment in stock into machines and equipment which are a rusting and wasting asset. Yields on which optimistic forecasts were based started to fall because of soil structure problems, an increased incidence of perennial weeds, and pests and diseases. An alarmed Ministry of Agriculture, isolated from farming by the bricks and mortar of Whitehall, and characteristically behind the times in its appreciation of the problems of the industry, set up a committee to report on this loss of yields and profitability which was aggravated by two very difficult seasons—a bad harvest followed by a catastrophic spring for clay-land farmers, who also had to suffer the effects of a severe summer drought.

Out of this, and the reporting committee could do little more than reiterate the sound premises on which alternate husbandry is based, came a realisation of the importance so far as a large area of lowland Britain is concerned of maintaining a sensible balance between pastures and crops. But pastures cannot be justified solely because of their contribution to succeeding tillage crops. They must, as the rental value of land rises, make their own positive contribution to farm profitability and herein lies the importance of a proper understanding of how to manage and exploit rotational grass.

CHEAP SOURCE OF NUTRIENTS

Fortunately in the economic climate that faces us, pasture, both temporary and permanent, is a relatively cheap source of nutrients. Even when barley was no more than £20 per tonne and high protein cake sold for about twice that figure, pasture was the cheapest source of nutrients for ruminant livestock, always providing a farmer was capable of managing his herd or his flock to make optimum use of these nutrients. Now inside the Common Market with a much less favourable relationship between concentrate costs and milk returns the situation is very different. Admittedly, the prices for beef and lamb have risen substantially, but neither beef cattle nor sheep are very efficient converters of expensive concentrate foods. Certainly so far as Britain is concerned, the future of its ruminant livestock production depends very much on this country's undoubted potential to produce grass.

In this respect it compares more than favourably with all the other countries of the Nine with the exception of the Atlantic seaboard of France and Eire. This latter country is in a highly advantageous position, as far as grassland farming is concerned, except for the congenital incapacity of the average Irish farmer to take life seriously. British farmers will only have real cause to fear the competition of Irish grassland when Dutchmen and Danes as well as Irishmen discover the new El Dorado that Ireland offers because of its substantial climatic advantages.

The truly continental areas—the heartland of France and Germany and most of Italy—are not well suited to pastoral production as we know it in Britain. Maize and lucerne, apart from wheat and barley, are the crops best suited to their climate and there is a big future for these once the Continentals resolve their farm structure problem. This is where we in Britain have an enormous advantage. Admittedly, we still have too many farms that are too small, in the context of a modern society that demands a herd of at least 50 cows or 500 ewes per labour unit to make an economic unit, but we also have a preponderance of the farmed area of Britain in units of sufficient size to give this country a competitive advantage. Additional to these natural and structural advantages over the Continental countries, there is the British flair for stock management. British farmers generally are good stock-men and this is especially true the further one goes north or west in these islands. One cannot come to any other conclusion than that British farming has a rosy future in the extended community by virtue of its grassland potential.

COMPLEMENTARY FUNCTION OF CONCENTRATES

Many grassland enthusiasts, and the senior author of this book has not been guiltless in this respect when he was younger and less temperate in his views, have extolled the virtues of pasture without a proper recognition of its limitations. These stem from two principal deficiencies; first, seasonal differences in productivity from mid-spring plenty to the winter trough, and second, the qualitative changes that occur over the grazing season. When pasture is at its best, in that leafy stage in early May, when it has the characteristics of "a watered concentrate", there is no immediate advantage to be derived from supplementation except perhaps to ensure an adequate intake of magnesium. But a very different situation obtains over the greater part of the year in Britain, especially in dairy production where a judicious use of concentrates will pay dividends.

Our situation in this respect differs very much from that in New Zealand, which is generally regarded as the world's leading pastoral country. Because of a favourable climate New Zealand has a very long growing season and farmers there have evolved efficient systems of seasonal milk and lamb production that are based entirely on pasture utilisation. Their returns from milk and lamb in relation to the cost of cereals are such that it simply does not pay to supplement pasture. Their example is valuable to British farmers in that they have a well-established confidence in the value of well-managed grass in achieving a high level of output per ha. This was something the average British farmer had to learn because up till a few years ago he remembered a time when concentrates were cheap, and fortified by advertising pressure from the provender trade, he continued to think of pasture in terms of maintenance plus 10 to 14 litres of milk. Even today there are tens of thousand of dairy farmers who continue to feed appreciable quantities of dairy cake in the month of May when grass is at its best. If they get as much as an additional litre of milk for every kilogram of supplement fed, they will be getting a better than average return but they won't be making any profit.

So far as dairying is concerned, the aim must be to provide high quality grass over as long a grazing season as is possible under the natural conditions of a farm. One must aim to conserve surplus grass before it deteriorates nutritionally and also to adopt methods of conservation that minimise nutritive losses. But having done all these things, there will still be a need, especially with an autumn calving herd, to use concentrates judiciously in order to get the best out of one's grassland. The position will be different with a spring calving herd or with fat lamb or beef production, but here again a sensible use of supplements can pay dividends. This is a point we will return to later in this book.

PRODUCTION AND PROFIT

When considering production, whether it is milk or meat, it is unwise to think only of individual yields because the law of diminishing returns applies just as much in the feeding of livestock as it does in the fertilising of crops. There are two main components that contribute to livestock output per unit area under grazing conditions. The first is the individual performance of animals in terms of milk production or liveweight gain and the second is stocking intensity. There is a conflict between these two components in that beyond a certain critical intensity of stocking, individual production tends to fall. However, it is well established that maximum yield per ha is not coincident with maximum production per animal. In other words, in the interests of a fuller and a more economic use of pasture nutrients, we can afford to sacrifice some measure of individual yields when we increase stocking intensity to its optimum level. For instance, under a system of set stocking we may produce 18 fat lambs per hectare, weighing 22 kg dressed weight apiece, but with forward creep grazing on the same area we could produce thirty lambs each weighing 18 kg. The management problem is always one of achieving the best relationship between stocking intensity and individual output to achieve the optimum economic results.

It is this sort of situation that makes grassland farming such an intriguing study. Obviously no good farmer likes to lose out on individual output if he can possibly avoid doing so, and the general inclination is to do everything one can to make the best of two worlds. If a farmer is sensible, he gradually increases his stocking intensity but then, to counter losses in individual productivity, he adopts all those measures that safeguard individual productivity—higher levels of fertilisation to increase output, better control of grazing to ration supplies and safeguard quality of herbage, better conservation methods, a more judicious use of supplements and so on.

There is no better spur to a more efficient utilisation of pastures than having a hungry herd or flock breathing down one's neck because when you are in this situation, and you have a conscience as well as an anxious bank manager, then you really start to do something about your pasture management.

CHAPTER II

CHOICE OF GRAZING ANIMAL

DAIRY cows are our most efficient grazing animals in the sense that they produce the greater quantity of saleable products and the greater monetary return per unit area in comparison with either sheep or beef cattle. It is primarily for this reason that dairying is practised on expensive land and especially on smaller farms. Indeed on a farm of less than 35 ha it is questionable whether the rearing of replacements is justified, for usually it is preferable either to buy point of calving heifers from a reliable source or, alternatively, to arrange for contract rearing of selected home-bred heifer calves.

There are several reasons for this comparative superiority of the dairy cow under circumstances that demand high intensity production. On the whole, meat production is a comparatively inefficient biological process when measured in terms of food conversion. Even under good management beef cattle are only about half as efficient in turning their food into human food as dairy cattle under comparable conditions, when one takes the contribution of calves as well as milk output into account. In

Britain, as well as in other countries of Western Europe, where there is such a pressure on land resources, the contribution of dairy-bred calves, surplus to breeding needs, is becoming increasingly important in beef production because the overhead cost of producing these calves is carried by milk production. It is for this reason that a truly dual-purpose breed like the Friesian is so important in Britain.

Though there is generally less capital invested in a beef-breeding cow as compared with a dairy cow, and though the latter has a greater food requirement, nevertheless the total value of production of the dairy cow, taking both milk and calf into account, is far in excess of that obtained from the suckler cow which has only two products—its calf and the headage subsidies which are bound to disappear in the Common Market. Admittedly, the labour input, with its double daily grind of milking, is very much higher with dairy cattle than it is with sucklers, but on small family farms this is not a serious economic restraint because a farmer is usually prepared and able to put time into his dairy enterprise since it gives him a much better living than any alternative use of his land.

Even on large dairy farms, with a mainly hired staff, the labour cost per litre has not kept pace with the rise in wages because of the quite remarkable increase in labour efficiency that has characterised the British dairy industry over the past two decades. That long succession of innovations—quick milking, circulation cleaning, bulk collection of milk, mechanisation of conservation, self-feeding of silage, cubicle housing and mechanical handling of slurry, to mention some of the more important developments—have made an enormous difference to labour efficiency and unquestionably this has contributed to the stability of the dairy industry.

DAIRYING MORE ADVANTAGEOUS

Another very important advantage of dairying over meat

production, which stems from the relative efficiency of dairy cattle, lies in the greater scope that a dairy farmer has for using both fertilisers and purchased concentrates. The use of concentrates with beef cattle or sheep must be sparing, with existing price relationships and limited to periods of critical importance; for instance, late pregnancy and the early stages of lactation in the breeding ewe or the final fattening stage with bullocks where a grain supplement is an essential addition to a mainly silage diet if the desired rate of gain and degree of finish are to be attained. Certainly in winter milk production, no matter how good he is in respect of the quality of conserved grass, no farmer can expect to get optimum economic performance from his herd without some feeding of concentrates. Apart from their direct contribution to yields they also have the effect of increasing the total carrying capacity of a farm. In a sense, when a dairy farmer buys concentrates he is adding hectares to his farm and in the process, provided these concentrates are used to good effect, he is achieving economies of scale.

NITROGEN USAGE

The same applies in the use of fertilisers, particularly nitrogen. If a farmer relies entirely on lime, phosphate and potash to maintain a balanced clover-grass sward, he will be fortunate to achieve a dry-matter output of 6,500 kg per ha. The application of 350 kg of elemental nitrogen, provided there is sufficient moisture, can boost dry-matter production to 11,000 kg or more per ha. But this boost in production and the utilisation of this additional nutriment costs money. Apart from the actual cost of the fertiliser, there is also the need for closer subdivision and more sophisticated control of grazing as well as improved standards of conservation.

The greater efficiency of the dairy cow can justify these costs but there is no evidence that this is the case either with suckler cows or stores that are being finished on pasture. The latter are slow to finish on grass-dominant swards that are the inevitable result of high nitrogen usage. Their best performance is generally on swards with a considerable clover content.

Generally, one can conclude that apart from dressing pastures for enhanced conservation crops, and possibly for early bite and back-end grazing, the feeder and breeder of beef cattle has a very limited scope for using nitrogen. The situation, however, is different in the eighteen-month system of beef production where autumn-born calves at a stage when they are relatively efficient food converters are growing frame during their summer on grass. Here there is considerable justification for high nitrogen usage, combined with controlled grazing, and the conservation of grass surplus to immediate grazing needs. With this system one must think in terms of about four beasts to a hectare of grass, and this level of stocking is not feasible unless one is using at least 200 kg of elemental nitrogen per ha each year.

Conventional fat lamb production, based on a summer stocking intensity of 10–12 ewes per ha plus their lambs, and some cattle grazing to control surplus grass, does not in our experience justify a heavy use of nitrogenous fertilisers. Again, as with feeding cattle, there is a better finish on the lambs if there is a strong clover element in the sward, but the situation is rather different under a system of forward creep grazing where one is able to run 20–25 ewes plus lambs per ha. Here the aim is not lambs fat off their mothers, except in the case of singles, but forward store lambs for subsequent finishing, or possibly for immediate sale when we really start to export lambs to Continental Europe, which requires a leaner carcass than that in demand in Britain. Fourteen years of experience with forward creep grazing on the University of Newcastle's Nafferton Farm indicate that the opportunist use of nitrogen to preserve a good sequence of grazing over the summer is an economic proposition. Certainly, an average summer output in excess of 900 kg liveweight per ha could not have been obtained if there had been a complete reliance on clover nitrogen.

One further advantage of dairy production over both beef and fat lamb production must be mentioned and this relates to cash flow. Admittedly, the dairy farmer has a much higher investment per ha in the form of livestock, buildings and equipment and there are also larger outgoings for labour, fertilisers and feedingstuffs, but there is also the monthly milk cheque and often the now substantial return from surplus calves sold at the week-old stage. The suckler man has to wait nine months for his annual cheque if he sells his calves as weaners, and if he elects to finish them as young beef, there will be at least six further months before the money goes into his bank account. The situation of the eighteen-month beef producer, buying week-old calves, is even more difficult because he has to buy another crop of calves and incur substantial early rearing costs before the first bunch has been sold. At its peak the total investment per ha for an intensive eighteen-month system can approach that of dairying.

RELATIVE ADVANTAGES OF SHEEP AND CATTLE

Normally in Britain one finds sheep in association with cattle rather than as a sole enterprise. There are several reasons for this. On upland farms, traditionally used for sheep production, there has been a move to cattle because Government policy over the past twenty years has, through the medium of preferential headage subsidies, been directed towards an expansion of beef production. By comparison, up till 1968 sheep have been Cinderellas.

On lowland farms there is age-old belief that under high intensity stocking a sheep's worst enemy is another sheep and that it is important to dilute the grazing hazards, particularly those arising from parasitic worms, by bringing in another specie of grazing animal. There is good evidence from many sources to show that mixed stocking does, in fact, reduce parasitic infection of lambs so that the widespread practice of associating cattle with fat lamb production has real justification if other more positive methods of controlling worm infections are not adopted.

There have been considerable advances in this connection in recent years as a result of the availability of more efficient drugs and a better understanding of the life cycle of the more important parasitic worms. Through a combination of strategic dosing and field hygiene it is possible to maintain a high intensity of stocking within a season on the same field with a negligible worm problem, provided that the field in question has no carried ewes and lambs in the previous year. The safest fields in this respect are maiden seeds but the same effect can be obtained by conservation and cattle grazing in alternate years as a means of disinfecting pastures that have been subjected to heavy sheep stocking. It is important, however, to drench ewes against the "spring rise" in worm egg output before they go on to clean pasture because undrenched ewes can lay down a significant level of infection that will impair late summer thrift in their lambs.

The possibilities of a greater intensity of fat lamb production, using more fecund ewes, have been greatly enhanced by these developments. With the growth in demand for meat in the world at large, and particularly in Western Europe, and the severe imitations on any great expansion in breeding cattle numbers, really intensive fat lamb production could have a new dimension in Britain which, apart from Ireland, is the only country in the Community that has any effective organisation for such production.

SHEEP AND CATTLE TOGETHER?

To return to sheep and cattle in combination, there is a widespread belief that when the two are run together there is a fuller use of grass and a greater production per ha than obtains when either species is grazed separately. Cockle Park supplied some evidence many years ago that there is greater production from sheep and cattle than there is from sheep alone. This conclusion was not based on any planned experimentation but from observations of performance on two adjacent fields. Anyone knowing the land in question and the rates of stocking that were adopted would be bound to conclude that there was a considerable measure of under-utilisation, particularly on the sheep alone enclosures which were under-stocked.

More comprehensive trials at Ruakura in New Zealand comparing breeding ewes alone, cattle alone, and ewes and cattle in combination, under conditions of full utilisation revealed no appreciable differences in meat production per ha. The expectations are that if this trial were repeated in Britain, and the necessary precautions were taken to control disease, the sheep alone treatment would be the more productive and in the absence of any discriminating subsidies the more profitable.

There are several reasons for advancing this view. The first is that British lowland ewes, which are mainly first crosses with a Border Leicester male parent, are much more fecund and deeper milkers than the Romneys used in the Ruakura study, which would not produce more than 18 kg of dressed lamb per ewe as compared with at least 27 kg for a British cross-bred ewe. Secondly, longer and more severe winters in Britain impose a much greater strain on cattle as compared with New Zealand conditions. It is admitted that British ewes suffer also a harder winter than their counterparts in New Zealand, but sheep as a species have a much better adaptation to the ups and downs of pasture production over the course of the year.

This especially is true of breeding ewes because maximum food demands coincide with the greatest abundance of pasture during the late spring and summer weaning, and the subsequent sale of lambs or their removal to feeding crops like rape and soft turnips takes a burden off pastures as they start to fail, while the normal autumn flush provides that rise in the plane of nutrition immediately prior to mating, which is so important in securing a good lamb crop. It is only in that period, immediately prior to and just after lambing, when better feeding is essential for these highly prolific ewes.

Add to the argument these points—there is the wool byproduct which despite the competition of synthetic fibres still makes a significant contribution to farm income. Then there is a better cash flow from the ewe flock than from a herd of breeding cows. In addition, there is an appreciably lower investment per ha in breeding stock than there is with a herd of sucklers.

BEEF UP, SHEEP DOWN

Why, then, in the face of these advantages for sheep has the past decade witnessed a spectacular increase in the size of the national beef breeding herd while the national sheep flock has shown a recession? The principal reason for this has been the relatively higher guaranteed prices that have been offered for beef as compared with lamb, while the preferential headage subsidies that beef cattle, both breeding cows and weaned calves, have enjoyed have also played an important part. The only headage subsidy received for sheep is that for breeding ewes on upland farms and this on a pro rata basis, measured in terms of livestock equivalents, has been very much higher for cows than for ewes.

Also, lamb and mutton in Britain has had to face much more competition from imports, principally from New Zealand, than has been the case for beef as a consequence of a reduced international trade in this latter commodity. It is true that there is a world shortage of beef and this is likely to continue. Because it is a preferred type of meat it is unlikely that there will be any slackening of demand but lamb, provided it is not over-fat, is also a meat with a strong potential demand, particularly in those Continental countries where it is now a luxury food. When one sees the prices being paid for what we could consider to be miserable undersized lambs in the south of France or in Northern Italy then one realises the tremendous market potential that exists for British sheep farmers in the expanded Community, especially when New Zealand supplies are limited by tariffs and the direct subsidies for domestic beef production are absorbed in the market return.

UPLAND FARM EXPANSION

Sheep production, especially on the more marginal land, could well have a very different appeal from that it has experienced in recent years. It is not suggested that this expansion will necessarily be to the detriment of suckler production. Not only is the demand for beef likely to be so strong that the loss of headage subsidies will be compensated in part by higher market returns, but also there are distinct management advantages in combining sheep and cattle under these conditions, partly stemming from the control of parasitic worms that has already been discussed but, even more important, from the improvement in the quality of sheep grazing that results from the judicious use of cattle where it is not feasible to control pastures by mechanical means as one can on the lowlands. There are many examples in Britain of upland farms that have been greatly improved by the introduction of cattle, and one of the great and sometimes unrealised benefits of the Government's policy of getting more beef from the hills has been the vast improvement that has been effected in the quality of upland grazings over the course of the past twenty years.

One last point about the production of suckler calves from upland farms, with the removal of headage subsidies farmers will become fully dependent on market realisations. The neat compact calf out of a Blue Grey cow by an Aberdeen Angus bull will no longer be a proposition even though it produces very high quality beef. The simple fact is that it has insufficient growth potential. When cows are mated solely for the production of calves for beef the expectation must be male animals that weigh at least 500 kg at 15 months of age and not tiddlers of 350–400 kg. The need on upland farms is a milky cow with frame, possibly a Simmental cross, that is mated to a bull with a high growth potential. There does not appear to be a breed superior to the Charolais for this top crossing function and it is only a question of time, as Charolais bulls become more plentiful and less expensive, that their use will become more widespread even under upland conditions.

PLATE 1a

Cattle can be useful tools in maintaining good sheep grazing by controlling rough growth.

PLATE 1b

Cross Hereford cows grazing in semi-hill conditions.

ICI Photo.

PLATE 2a
Precision chop harvester in action. The silage is being treated with an additive contained in a 200 litre drum.

ICI Photo.

PLATE 2b

Eighteen-month beef cattle towards the end of their first summer at grass.

ICI Photo.

Grass farming is much more than the creation of productive swards. Grass nutrients must be turned as economically as possible into saleable products and the genetic quality of stock, be they beef or dairy cattle or sheep, and the quality of the management to which they are subjected are key factors in achieving this end.

CHAPTER III

LEYS OR PERMANENT PASTURE

"AN abundance of fresh food is not compatible with a superabundance of permanent grass."

Sir George Stapledon made this statement at a time when pastures were neglected. He was expressing his profound belief in a system of farming where the plough was taken regularly round the farm to bring vigour to swards based on selected species and varieties rather than on volunteer plants of doubtful quality. The surveys that he and the late Dr. William Davies had organised before the war revealed that an overwhelming proportion of the eight million hectares of permanent grass existing at that time consisted of bent grasses and had a grave deficiency of clover.

The translation of Stapledon's philosophies into farming practice was of enormous importance to Britain, especially during the war years. Some three million hectares of this poor pasture land, and large areas of rough grazings in addition, were brought under the plough, and most of it has been retained in alternate husbandry.

The whole process constituted a major revolution in British

farming. It represented more than a regeneration of pastures. It brought about a revitalisation of the whole industry, in that it engendered a flexibility of outlook and a spirit of adventure that had been lacking since the days of high farming a century ago. When these events become history they will have a significance similar to those of the eighteenth century that led to the spread of the Norfolk four-course rotation.

THE CASE FOR LEY FARMING

One cannot argue against the validity of the ley-farming concept under conditions of extreme pressure when food was rationed, for then it was virtually a question of food at any price. Nor can one suggest, under present conditions, a better system of land use for many of our farms which are above-average size in the main arable areas of Britain. Here ley farming is completely sensible, especially on the stronger soils where it is important to maintain good structure if crop yields are to be maintained.

In this context, one is thinking of those farms which are sufficiently large to carry the costs of mechanisation and which, as a group, are able to produce cereals as efficiently as anywhere in the world. Cash crops are the first consideration on such farms and, in a sense, output from leys in the form of milk and meat is a by-product of tillage farming, for the primary function of the leys is to put heart into the land. Properly organised, the associated livestock enterprise can make an appreciable contribution to farm income, and the whole constitutes a well-integrated system of farming which has considerable resiliency in meeting the contingencies of seasons and prices.

The purpose and value of leys are very different, however, on farms where the principal source of income comes from grazing animals. Though this may seem like heresy, in that it cuts right across the Stapledon philosophy, the arguments in favour of very long duration pastures and good permanent grass, as opposed to short leys, are difficult to refute on a large number of our farms.

This is especially true for farms of below-average size which are not suited to intensive cash cropping, and for farms of any size in the high-rainfall areas, especially on heavy soils where cropping is hazardous because of weather conditions at sowing and harvest.

Furthermore, the force of these arguments grows stronger as our farming passes into an economic climate where the emphasis is less on volume of production than it is on economy of production. The basic reason for this statement is that the cheapest nutriment we can provide for stock is that from productive permanent pastures. Expressed in another way, a system of pastoral dairying, fat lamb production or cattle feeding based on permanent grass has a much lower potential cost structure than one that carries the cost of tillage implements and the expenses of renewing pastures every few years.

EXPERIMENTAL EVIDENCE

Protagonists of ley farming will immediately answer that it is impossible to get the level of output from permanent grass that one can get from leys, and so scale of operations will be reduced. Instead of milking 40 cows, or running 250 ewes, on a given area of grassland, a farmer will have to cut his carrying capacity to such an extent that any economies he will make by swinging to permanent grass will be more than accounted for by losses of income due to a scaling down of enterprises. After all, they will say, there is the evidence of those comparisons between leys and permanent grass initiated by the R.A.S.E. and reported by William Davies and T. E. Williams. They concluded that if the permanent pastures of this country were replaced by well-managed leys there would be a fifty per cent increase in productivity.

There are two points which must be raised in answer to this. The first is that level of production is not synonymous with level of profit. The second is that a comparison is not valid if it is between well-managed leys and badly-managed permanent grass. The two types of grassland must have the same kind of management, and one would hope that it is an enlightened management which gives both types a chance to show their capacity.

It is important to note that in the R.A.S.E. trials, where a comparison was made between a ley and a first-class permanent pasture in the Welland Valley, which is famed for the quality of its permanent grass, there was no difference in production from the two types of pasture. Where the comparisons were made between leys and poor permanent pasture, then a fifty-per-cent advantage could be attributed to leys. In our context we are not concerned with this kind of permanent pasture, but with the superior sorts that can be created by intelligent management.

A few years ago at Cockle Park there was a comparison between permanent grass and three different leys based on perennial ryegrass and white clover. All four swards received identical grazing management and fertiliser treatments. At the end of three years the permanent grass showed a slightly higher output than the leys, and generally through the trial it had a much better performance during periods of drought. At any time during the growing season the permanent grass had more sheep-keep on it than any of the leys, and it had no establishment cost.

A long-term comparison has been made between permanent grass and leys on the Ministry of Agriculture Husbandry Farm in Lancashire, which is in a difficult farming area. It has revealed no advantage in using leys, and indeed results suggest that permanent grass may be the more economic proposition under these conditions of farming.

This comparison, almost more than any other, has made advisory officers stop short in their tracks and ask themselves whether their advocacies, based on an unquestioning acceptance of the Stapledon doctrine of ploughing and reseeding at regular intervals, are sound on farms where the purpose of grass is to feed stock rather than put heart into tillage land. Some of the more thoughtful have been debating whether they might not have been doing more good persuading farmers to improve their permanent grass by surface treatment rather than by ploughing, for there are still five million hectares of permanent grass in this country which have survived the war-time ploughing-up orders and the lure of subsidies that were offered for breaking old grass.

FLEXIBILITY OF LEYS

In any direct comparison of leys and permanent grass it must be said in favour of leys that they give much more flexibility in grassland management. Young leys, especially those that are based on early- and late-growing species, have a much longer growing season than permanent grass, which is generally regarded as a middle-of-the-season producer. This is certainly the case if the permanent pasture has a high proportion of bents or *Poa trivialis,* but if it is so managed that perennial ryegrass is the principal component, its spread of production does not differ markedly from an established perennial ryegrass ley.

In New Zealand, where the dairy industry is almost entirely based on permanent grass, a technique has been developed of over-sowing permanent pastures with an early-growing ryegrass. This has been remarkably successful in extending the availability of grazing without any need to plough.

Today with the availability of efficient forage harvesters and the development of self-feeding or easy-feeding of silage, the pressure for out-of-season grass is not quite so great as it used to be. Sometimes one wonders whether we would not be more advantageously engaged growing more grass when conditions are favourable rather than complicating life in a struggle to grow out-of-season grass. By adopting such a policy, we could have vertical grazing at the silage face for five months of the year, and seven months' horizontal grazing on pasture for the remainder of the year. It would, of course, be essential to have one's silage making well organised so that a first-class product is made from the grass that is surplus to immediate grazing needs.

Another argument in support of ploughing and reseeding is a belief that a pasture is most productive in its first harvest year, and that there is a progressive deterioration from then onwards. The senior author was given a very different viewpoint as a young man in New Zealand, namely that it takes several years following establishment to get a pasture into full heart. The process, we were told, was one of applying phosphate, potash and lime, where these were necessary, to get a strong establishment of clover. Then, by heavy stocking, with a consequential heavy return of stock excrements, to get a mobilisation of fertility in the top few cm of the soil. This would support a vigorous growth of the high-producing species which would crowd out the unwanted volunteers that only gain ascendancy if fertility is depleted. There are millions of hectares of really productive permanent grass in New Zealand which provide support for this view.

Good management maintains the conditions that are essential for the dominance of sown species. If a ley deteriorates, one may be reasonably certain that some essential elements are lacking from the soil, or that there has been some form of mismanagement to tip the balance in favour of unwanted species.

In this last connection, one has to be a realist. Despite the best of intentions, the exigencies of farming may make it necessary to mismanage a pasture so that it deteriorates to a point where its restoration may be more quickly achieved by ploughing and reseeding than by any other means. Under such circumstances it may be sensible to plough, but this is a very different approach from a doctrinaire view that the plough should be put into a pasture every five or six years, whether it needs it or not. There is no need for every pasture on a farm to become a sacrifice area. Once a pasture has been established on a grassland farm every reasonable effort should be made to keep it in full vigour so that its establishment costs are spread over as many years as possible, even to the point where it can truly be described as permanent grass.

VIRTUES OF PERMANENT GRASS

Turning now to the intrinsic virtues of permanent grass, other than its cheapness, the first is its capacity to recover from poaching, which is far superior to that of a ley. This virtue is almost a handicap for permanent pastures, because most farmers, recognising this advantage, will punish an old pasture in the early spring. They would be surprised at the greater productivity obtainable from their permanent grass if it was nursed as carefully as their leys and given the same fertiliser treatment.

A second virtue is its relative safeness so far as metabolic diseases and bloat are concerned. Old pasture, with its greater variety of species, is less incriminated in this connection than the simple quick-growing ley with grass all at the same stage of maturity.

A compensating disability is that a permanent pasture, under continuous stocking with the same species, may have a buildup of infections, for example, intestinal and lung worms. But there is no reason to believe that the clean-field technique (resting in alternate years from ewe and lamb grazing) is not just as effective with permanent pasture as it is with leys.

A further virtue of permanent grass is the way it stands up to summer drought, and its capacity to maintain milk yields in June and July when leys are tending to run to head. At this time of the year permanent grass carries much more leaf than leys, and invariably when a herd moves from an established ley to good permanent grass, there is a lift in production.

IMPORTANCE OF A PERSPECTIVE

The issue between leys and permanent grass, however, is not a black and white one, for there are so many shades of grey, according to the conditions on individual farms. The wisest course for many farmers will be a combination of leys of varying duration with permanent grass. The leys will give that extra measure of flexibility in the provision of early grazing so often missing if one has to rely on permanent grass alone.

The need for this flexibility will be greater on dairy farms than on lamb and beef producing farms, and this is fortunate, because dairying is better able to carry the higher costs of ley farming. At the same time, there is also a great deal to be said in favour of a high proportion of well-managed permanent grass on dairy farms not large enough to justify the high capitalisation involved in the ownership of a wide range of tillage implements.

One of the most serious criticisms one can make of the ley-farming doctrine is that it has encouraged the small man to make his farm a replica of the big farm, which is able to carry the cost of machinery and which has the elbow room to grow corn and provide bedding straw. Would it not be more profitable, in many instances, for someone with a 25 ha farm to have all but, say, 4 ha either in well-managed permanent grass or very long leys? These four hectares could be devoted to a rotation of Italian ryegrass and late-sown kale, with two hectares of Italian being seeded down every spring. All straw, concentrated foods, and possibly even hay, could be bought in from someone able to produce these things more economically. Thus the only field equipment needed would be a tractor, a tipping trailer, a small forage harvester, and a spinner top-dresser. Such cultivations as were needed for the small arable area could be done by contract.

A farm organised in this way would have a low investment in machinery and would be straightforward to run, for there are no complications, and, with intensive management, it could have a milk output of at least 225,000 litres a year.

Such a plan of farming would, of course, demand a high level of management for permanent pasture. If you are happy to accept only what nature offers, which on all but the most fertile land will be daisies, buttercups and bents, then it is wise to forget all that has been said in favour of permanent pasture. If, on the other hand, you are prepared to provide the necessary fertility and management, you will find that the difference in production between leys and permanent pasture is surprisingly small, while cost of nutrients will be in favour of the permanent grass.

CHAPTER IV

MANAGEMENT AND PASTURE COMPOSITION

AN association of pasture plants is never static, because the balance of the component species is constantly changing in response to a wide variety of stimuli. Some of these changes are progressive, for example, the steady deterioration that is so often observed in sown pastures due to an invasion of inferior species as a result of mismanagement or failure to provide the necessary fertility to support the sown species. Conversely, though unfortunately less often, one sees the reverse process, where a superior combination of species is created by appropriate management of poor pastures.

Other changes in pasture composition are seasonal in nature, for example, the greater dominance of grasses in the early spring and autumn over clovers, which normally make their greatest contribution in the late spring and summer.

The aim of pasture management must be to maintain the most desirable combination of species by the various means which are under a farmer's control. This would be a relatively straightforward proposition if the appearance of the pasture were the only object, but a further over-riding consideration is the basic needs of stock. Sometimes the requirements of animal and of pasture are in conflict, and there has to be sacrifice in one direction or the other to obtain the best economic result. Such decisions require very fine judgment, and herein lies the art of being a good grassland farmer.

There are very few definite rules that can be followed except that stock should not be made to suffer at critical times, for example, lactating cows or ewes, or young animals when they are making most active growth. Ewes and single-suckling cows, however, have low nutritional requirements when their offspring have been weaned, and can be made to work for their living in the interests of pasture management.

Similarly, there are critical stages in the life of pastures when they should not be abused. This is especially true in the early spring, when they are particularly vulnerable to poaching and to the ill effects of over-grazing; the younger a pasture is, the more liable it is to suffer from such maltreatment. This exemplifies one of the advantages of having some permanent pasture on a farm, because it has greater recuperative powers from poaching. In the absence of such land or, better still, housing to keep horned stock off pastures, it is preferable to work to a system of "sacrifice fields" which are intended for immediate ploughing. In this way damage can be kept under control and the pastures which are intended to carry the burden of grazing or conservation in the subsequent year can be nursed in the early spring.

FACTORS DETERMINING COMPOSITION

When a pasture has been established there are three main sets of factors which determine pasture composition. They are:

(a) the physical condition of the soil
(b) soil fertility and fertiliser practices
(c) the timing and intensity of grazing and cutting.

There is a fourth factor which operates at the point of establishment, and this is the choice of seeds which are sown. The dice is loaded against the survival of sown species if non-persistent varieties are sown, for example commercial grades of so-called perennial ryegrass which have a high proportion of pseudo-perennial seed. Though this kind of seed is still on the market and is sometimes incorporated in seeds mixtures, one of the great advances of the last thirty years has been the increased reliability of herbage seeds through the development of certification schemes.

Today there are abundant supplies of certified pasture seeds, either bred strains or genuine old pasture ecotypes, and though they cost slightly more per kilogram than commercial seeds, they are no more expensive in the ultimate because one can safely adopt lower seeding rates. The net result is that their virtues of persistency cost nothing. Anyone sowing only non-persistent commercial varieties in a pasture which is intended to have more than two years' duration should have his head examined, because his is being penny wise and pound foolish.

PHYSICAL CONDITIONS

Physical condition of a soil is only partly under the control of a farmer. He has to accept the inherent features of any given soil type so far as they are determined by such factors as aspect, and proportions of clay, silt, and sand. There are, however, two important farming operations which can have a considerable ameliorating effect on difficult soils. These are liming and drainage.

Lime has more than physical effects, because it corrects acidity and increases the availability of certain other plant foods. It has a profound effect on structure, and in this way improves soil moisture relationships. Lime is not expensive in Britian so there is no excuse for lack of lime being a limiting factor, except where considerations such as contour make its application prohibitively costly.

The benefits of drainage, so far as pastures are concerned, are three-fold. The first arises from the fact that high-producing species will not persist under wet conditions, which favour an invasion of poorer grasses and useless weeds like rushes and buttercups. The second is an extension of the growing season, because wet land is also cold land. The third is that one has greater freedom in getting on to land to graze it, or for cultural operations like top-dressing or mowing. Though a farmer may not be able to face the cost of draining his whole farm adequately at one fell swoop, he should at least have a progressive programme of farm drainage to work to if surplus soil water is a problem. Otherwise he cannot hope to realise the potential of his grassland.

SOIL FERTILITY

The importance of providing adequate soil fertility, not only to give the right environment for the persistence of desired species, but to ensure that they are fully productive, should require little stressing. Yet the unfortunate fact is that many farmers are still much too parsimonious in their grassland fertilising practices. On too many farms the decision to plough a ley is made not because the land is needed for tillage but because the ley has "run out", even though management, apart from fertilising, has been good.

Generally the principal deficiency is phosphate, and without adequate phosphate clover will be sparse or virtually non-existent. This was the great lesson of Somerville's classic work at Cockle Park. A pasture without clover is only at half-cock, unless one is prepared to use really massive dressings of nitrogen. If there is not sufficient available nitrogen in the soil, high-producing, nitrogen-demanding species like ryegrass will disappear, to be succeeded by low-producing bents which can tolerate nitrogen deficiency.

Potash is another vital plant food which is likely to be

deficient, not only on chalk soils and gravels which are naturally poor in this respect, but even on heavier land which has been used continuously for conservation purposes. Every tonne of grass dry matter contains approximately the equivalent of 37 kg of muriate of potash, and unless remedial top-dressing follows cutting, such land can quickly become potash deficient. This is reflected in a loss of vigour in clovers, which in cases of severe potash deficiency have characteristic brown markings on the edge of their leaves.

Another element which is essential to clover growth is molybdenum, but is required only in very small quantities. Molybdenum deficiency has not been established on any scale in Britain, but there are parts of Australia and New Zealand where the application of as little as 150 gms of molybdenum salts per ha makes all the difference in clover establishment.

Of the major elements, nitrogen has a profound effect on the balance of grasses and clovers in a sward. When land cropped with a long succession of exhaustive tillage crops, which have depleted soil nitrogen, is laid down to pasture, there is a very strong growth of clover in the first year. This especially is true on light soils like chalks and gravels, if there are adequate phosphate and potash reserves.

Gradually, however, the grass component achieves ascendancy as the nitrogen status of the soil is replenished by fixation of clovers, with the sub-surface transference of nitrogen being augmented by nitrogen returned in the excrements from grazing animals. This latter contribution can be a substantial one, amounting in one trial with only an average sort of pasture to the equivalent of a tonne of sulphate of ammonia per hectare.

In a sense, clover is self-destructive, for it encourages grass as it builds up the nitrogen status of soil to a point where there is a balance with these companion grasses. The nature of this balance will, of course, be greatly influenced by the management which is adopted. This, as later paragraphs will show, profoundly influences the proportion of clover in a sward.

The application of fertiliser nitrogen, as one would expect

from the foregoing, has a considerable effect on the balance of clover in a sward. Even quite light dressings of, say, 50 kg early in the season will reduce appreciably the proportion of clover in the sward during the subsequent summer. Heavy continued dressings of the order of 200 kg N per ha annually will virtually eliminate clover if it is associated with ryegrass. The primary cause appears to be the stronger competitive power of grasses and the effects of smother. If the pasture is fairly tightly grazed the loss of clover is not so great as it is if the grass is allowed to get away. There may be a secondary effect with supplement ammonium nitrate if special efforts are not made to combat the rise in soil acidity which results from the continued heavy use of this fertiliser.

It will be realised from this that there is something of a conflict between clover and fertiliser nitrogen. This is a point of great consequence in grassland farming, because there is a danger of falling between two stools. This will be debated at greater length in the next chapter, when an attempt will be made to resolve the issue as it affects different farming circumstances.

CUTTING AND GRAZING

Turning now to the effects of cutting and grazing management, the classic work in this field was that undertaken by Professor Martin Jones when he was at Jealott's Hill, some forty years ago. Starting with a uniform pasture, he was able to create very diverse types by variations in timing and intensity of cutting and grazing. On the one hand, he created highly-productive ryegrass and white clover pastures, while on the other he allowed the processes of deterioration to move to the point where weeds and inferior grasses dominated.

This latter sward, which was the victim of overgrazing in the early spring and undergrazing in the summer, was typical of large areas of British grassland at that time. Unfortunately we have not yet seen the end of this sort of grassland, if the views provided by British Railways in the summer months can be taken as a fair indication of the state of our pasture management.

Martin Jones' work, and that of others who have developed our understanding of grassland ecology, have made it possible to state these fairly definite principles on which to base the handling of pastures:

(a) Hard grazing at a time when one species is making very active growth tends to put that species at a disadvantage relative to a companion species with a different growth rhythm. This is well illustrated by the usual management of grazing swards which are intended for the production of wild white clover, for example in Kent. Here ryegrass/white clover swards are hard grazed in the spring when the ryegrass is growing rapidly, but stock are removed in June when, in all but the wettest of seasons, the clover will achieve dominance. If the intention is to take ryegrass rather than clover seed, fields should be closed early in April following a top-dressing with nitrogen.

(b) Species vary in their response to light and shade, for example, the extreme pasture types of ryegrass and white clover require light if they are to compete successfully with a tall-growing, shade-enduring species like cocksfoot. Hard grazing of a Cockle Park mixture, especially with sheep, will soon result in what is virtually a pure ryegrass/white clover sward. Repeated annual cutting at an advanced stage of growth, or even continued lax grazing, will result in cocksfoot dominance and the almost complete disappearance of clover. Cocksfoot dominance becomes particularly marked if a sward is laid up for foggage for several years in succession. The same treatment encourages Yorkshire fog in permanent pastures.

(c) Mechanical damage, which is likely to occur on wet soils in the early spring, will open up a sward and promote conditions favouring the spread of light-demanding weeds like buttercups and daisies. We often see this result on heavy soils in several parts of England, for instance on the Weald of Kent and in the Midlands, and especially on small dairy farms, where stock

PLATE 3a

Lime improves soil structure, corrects acidity and increases the availability of certain plant foods.

PLATE 3b

Applying nitrogen for an early bite. Timing of application of nitrogen dressings in spring is very important.

PLATE 4a
Good drainage encourages high-producing species, reduces weeds and extends the grazing season.

PLATE 4b
Welsh mountain draft ewes with Suffolk cross lambs at the Welsh Agricultural College's Tanygraig Farm. The intention should be to have lambs which are big enough in late April/May to make full use of the grass flush.

going on the land when it is still wet do more harm to the pastures than good for themselves.

(d) The maintanance of vegetative growth in pasture plants is dependent on the pruning of tillers just before the point at which they throw up seed heads. Even an annual like wheat can be kept growing for several years, as Martin Jones has shown, if the seed-bearing shoots are cut before they break head. Once an annual has produced seed it will die. A perennial, though it survives after seeding, goes into a period of relative dormancy—a situation which is commonly shown by the poor aftermath growth in a pasture which is cut for hay at a very mature state.

(e) Continuous hard grazing in the growing season tends to reduce total production of a sward and to discourage the more productive species, except under the most favourable conditions of soil fertility. One must recognise in this connection that grass leaves have another function apart from that of providing grazing for livestock. They are also vital to the parent plant for their part in the photosynthetic process which creates the complex substances like carbohydrates and proteins that livestock utilise.

Woodman, working at Cambridge in the late twenties, illustrated this point very convincingly with simple mowing experiments where he cut pasture at intervals of one, two, three and four weeks. There was a progressive and substantial increase in yield as the intervals between cutting increased.

LEAF AREA AND ROOT DEVELOPMENT

Australian research workers have shown the importance of a pasture plant retaining a reasonable leaf area if one is to get sustained production. They argue that continuous grazing, provided there is no over-grazing and a reasonable leaf area is maintained, will give more production than rotational grazing or fold grazing, where a pasture is allowed to get up and then is completely bared of leaf, especially if difficult environmental conditions follow. Undoubtedly this is true with fold grazing in the early spring if cold north-easterly conditions set in and the pasture is grazed right to the bone. Recovery will be very slow, though the ungrazed section of the field may still be making appreciable growth. Here it is possible that there is also a micro-climatic effect operating, for the grass cover could protect the ground against loss of heat.

In the early spring there is wisdom in adopting a policy of fairly lax rotational grazing of early-bite pastures, especially those containing clover, with the removal of stock from the field when there is leaf remaining to grow more leaf.

Another factor operating is the influence of defoliation on root development. Many years ago Jacques in New Zealand demonstrated the relationship of root development to top growth. Hard grazing reduces root development, and this may seriously impair production with the onset of dry weather unless the soil is fairly moist because of a high water table. Probably the latter factor is operating on Romney Marsh where, despite very close grazing, high production is achieved and a desirable pasture association preserved.

IMPORTANCE OF PASTURE CONTROL

It could be argued that some of the foregoing statements are conflicting, but one may fairly conclude that, for most conditions, it is wise to avoid overgrazing at those stages when pastures are making rapid vegetative growth. This may be achieved by lax continuous grazing or by controlled rotational grazing, without baring the pastures too hard so as to impair recovery.

The difficulty with lax continuous grazing is that the pasture tends to become a mosaic of short succulent growth which the stock prefer, and long mature growth which they neglect. Concentration of stock in large numbers for a short period gives more chance of preventing the mosaic developing and reducing the risk of weed invasion in the rough patches, but even here it may be necessary to use the mowing machine to prevent seed head emergence. The sensible thing to do, however, is not to mow for the sake of mowing but to integrate conservation with pasture control. Cutting grass for silage is more than a means of building up winter food reserves, for it is also a part of pasture management.

Later in the season, during July and August, it is often essential to eat or cut a pasture very bare for its own good. This point is reached when most grasses are undergoing their summer pause and there is a good deal of neglected rough growth which must be removed to prevent loss of clover. Here a flock of ewes which have had their lambs removed are an ideal tool to turn this rough growth into dung and urine. It will do the ewes a lot of good, too, because it is in a farmer's interests to have them on short commons at this time.

From all this it will be realised that timing and intensity of grazing can be varied to alter balance in mixed swards. For instance, if clover is weak one must endeavour to prevent shading in the spring, while cocksfoot dominance can be corrected by fairly close grazing throughout the year. A pasture is a very malleable complex of species, both sown and volunteer, but someone who has an understanding of the ecological principles involved is in a position to control pasture composition to suit his farming needs.

CHAPTER V

THE IMPORTANCE OF CLOVER

PROBABLY the greatest single step in the progress of British agriculture in the last three hundred years was the introduction of red clover. It made possible the much more productive system of farming which became known as the Norfolk four-course rotation. Tillage farmers learned very quickly the value of clover, not just as a forage but for enhancing the yields of succeeding crops.

White clover, the normal clover constituent of pasture, is not an introduced but an indigenous species, and no doubt it was an important component of grazing land on the more fertile soils in mediaeval times. Though it was included in sown pasture mixtures at least one-hundred-and-fifty years ago, it was not until the beginning of this century that its essential value in pastures was realised.

Gilchrist working at Cockle Park, and Findlay at Aberdeen, were the two men, above all others, who brought home to farmers the importance of a vigorous white clover in the ley, not just for its contribution to production, but in stimulating growth in the following cereal crop. Today we associate the name of Stapledon with ley farming, but he always recognised the vital part that these two men played in establishing the system.

The value of clover in a sward was brought home to the senior author as a student in 1932 by a simple experiment which compared the production of a pure ryegrass sward with one containing white clover. The latter was more than twice as productive, but analysis of the herbage showed that the difference was not accounted for by the clover fraction alone, because there was a fifty-per-cent greater yield of ryegrass where it was growing in association with clover. The stimulus came from the sub-surface transference of nitrogen which was fixed by bacteria in the clover-root nodules.

One New Zealand worker has put the nitrogen-fixing capacity of clover in a really good sward at a figure as high as 550 kg of applied N per ha per annum. The figure for an equivalent pasture under British conditions, with lower soil temperatures and a much shorter growing season, will be only a fraction of this, but nevertheless it will be substantial, and probably not less than 170 kg N per ha. The direct effect of clover nitrogen, of course, is augmented under grazing as opposed to mowing conditions. The animal retains only a small proportion of the nitrogen from its food, the balance being excreted in the dung and urine to make a further contribution to pasture growth.

The combined effect of sub-surface transference of nitrogen and excreted nitrogen was strikingly illustrated in an experiment which was conducted at Wye College a number of years ago. The trial constituted a comparison of the four white clovers–Dutch, New Zealand, S 100 and Kent. In the first year of the trial there was little to choose between the clovers, all swards producing approximately 800 kg liveweight increase per ha from sheep. At the end of this year, however, there was a distinct tailing off in production from the Dutch paddocks, in which there was an almost complete loss of this short-lived variety. This was reflected in its performance in the second year, when liveweight production fell to just over 550 kg per ha, while production from the other three clovers rose to the 900–1000 kg level.

These relative positions were retained in the third year. In the fourth year there was also a loss in production from the New Zealand and S 100 clovers, which fell behind the Kent. These two rather upright growing types of clover did not, in a year of drought, stand up to the intensive grazing as did the prostrate Kent clover, which was evolved under conditions of very close grazing.

The effects of these clovers were reflected in a following potato crop which was used to measure residual fertility. The highest yields were obtained from the plots which had carried the Kent clover, and the lowest yields from the former Dutch plots.

Returning to the grazing phase of the trial, in the second and third years one would have thought that the companion S 23 ryegrass, common to all the swards, was a completely different variety on the Dutch paddocks from the ryegrass on the other three treatments. It was yellow, stemmy and, according to the behaviour of the sheep, very unpalatable. They spent approximately an hour longer each day grazing, and though lower stocking rates were imposed, there was no finish on the sheep. One could not have wished for a more graphic and convincing demonstration of the importance of sowing a reliable white clover and adopting methods of management which ensure its persistence.

LIMITATIONS OF CLOVER

It will be argued that no-one in his senses will ever sow a non-persistent white clover like Dutch in long leys. But this is beside the point, because for one reason or another there are millions of hectares of British grassland deficient in clover. As a consequence, they are incapable of producing to full capacity unless those concerned with their management are prepared to be heavy users of fertiliser nitrogen. We know from fertiliser usage surveys that this is not the case.

From a farm manager's viewpoint, clover nitrogen is cheap

nitrogen. It costs no more than the price of the original seed and the lime, phosphate and potash, which form a basic dressing whether one uses fertiliser nitrogen or not. Though it is true that under British conditions one cannot hope to obtain the levels of herbage yields, even from a well-managed clovery pasture, that are obtained from pastures which receive really heavy nitrogen dressings, nevertheless the cheapest nutriment is obtained from a reliance on clover. The upper limit of production in Britain will, however, be in the vicinity of 7,000 kg of dry matter per ha. With an annual application of 200–300 N kg per ha, it is reasonable to expect yields of 10,000–11,000 kg of dry matter, and even more under favourable conditions.

This brings us right into the business of grassland farming. Heavy nitrogen dressings are fully justified if (a) the cost of nitrogenous fertilisers is low relative to the returns for the animal products from grazing, and (b) if there is an efficient use of the additional herbage which is produced as the result of such dressings.

In New Zealand, where the sale value of milk is only about one-third of that in Britain and the cost of nitrogenous fertilisers is twice as high, it does not pay to apply nitrogen, even on the most efficiently managed dairy farms. Here economics dictate a reliance on clover nitrogen. It is, in fact, no very great hardship to do this. With the very long growing season, a farmer there can expect about double the dry-matter production we can expect in Britain when we rely on clover alone as a source of nitrogen.

FUNCTION OF FERTILISER NITROGEN

Our position constitutes a very different situation from that in New Zealand. If one reckons that a cow needs about 5,000 kg of grass dry matter annually in addition to concentrated food, about 0.8 ha of good pasture without nitrogen will be required to support a milking cow. With heavy nitrogen dressings, of the order of 300–350 kg per hectare, it is reasonable to run a cow to half a hectare of grassland.

Expressed in another way, if a farmer carries 40 cow equivalents on 32 ha of grassland without using nitrogen, an application of 350 kg N per ha will enable him to push his herd up to 70–80 cow equivalents and in the process increase his gross income by a least two-thirds. There will, of course, be a proportionate increase in variable costs such as those for purchased food, veterinary services and so on, but fixed costs such as those of management and rent will be spread over the much larger output. If work can be organised through a better layout and appropriate mechanisation, there need be no increase in labour costs, but there can be a very substantial increase in net income. This has been demonstrated by many successful dairy farmers.

Heavy nitrogen use in this way has a special appeal to the small dairy farmer, who is particularly handicapped by lack of land and an under-use of available labour. It is within his power to add the equivalent of at least fifty per cent to the area of his grassland, and though this will involve a considerable outlay per ha, he is able to increase net income because of the considerable economies of scale which can be effected.

Nowhere is this philosophy of grassland dairy farming seen to better effect than in the Netherlands, where the contribution of clover is virtually ignored because it does not permit the intensity of farming which is necessary for a livelihood on these small Dutch farms. While the cost of grass nutrients is higher than it would be if they relied on clovery pastures, nevertheless it is still considerably lower than it is for purchased foods, and there are the compensating economies from being able to keep more cows on a farm.

The Dutch example is being followed in the United Kingdom, especially in Northern Ireland, another country of small farms, which fortunately has the sort of climate that assures full value from applied nitrogen. However, many of our dairy farmers who are attempting to make good use of their grass tend, in a typically British fashion, to steer a middle course between the Dutch and

New Zealand extremes. There can be a danger of falling between two stools by such an action, because of a conflict that exists between clover and bag nitrogen at low rates of application. For instance, if one applies 70 kg N per ha in the early spring, there is an immediate encouragement of grass at the expense of clover. If no further nitrogen is applied, there is usually an appreciable slumping of subsequent production due to the impairment of clover contribution. The net result is that total annual production will be no more, and sometimes less, than it would have been if nitrogen had not been applied. The main advantage that this nitrogen gives is a better spread of grass production.

It can be a very sensible management decision to forego some May grass in order to cut down cake feeding in late March and early April, but it is questionable if the application of this sort of early dressing to the whole of the grassland on a farm is the right course. If one is steering a middle course and trying to get the best of both worlds, it may be better to be a Dutchman on an appropriate part of the farm and a New Zealander on the rest of the farm. One would apply the heavy nitrogen dressings to pastures such as those based on Italian ryegrass, which are particularly responsive to nitrogen, or old leys which are probably due for the plough because they have lost their clover content.

The above comments apply particularly to swards which are intended for grazing rather than cutting, and they are also more applicable to areas with a good summer rainfall and therefore more likely to have a stronger clover contribution. The case for using higher nitrogen applications is stronger in the more difficult grassland areas, in the north of the country, and on heavy soils. The occasional nitrogen dressing appears to be used more efficiently by a pasture that is allowed to grow to the cutting stage than one that is closely grazed.

Any advocacy of a heavy use of nitrogen must be qualified with the stipulation that, to make it pay, one must adopt really efficient methods of utilisation. It is not enough to grow fifty per cent more herbage by applying a given quantity of nitrogen over the course of the year. It must be turned into fifty per cent more milk or meat.

If there are surpluses beyond immediate grazing needs, these must be turned into hay or silage reserves so that any subsequent increase in carrying capacity can be made with confidence. If any efforts to increase the intensity of grassland farming by a greater use of nitrogen are to be successful, there will be a concomitant of more intensive stocking, and the conservation of surpluses in such a way that a first-class product results.

There may be some latitude for being a little casual with grass dry matter that costs £12 per tonne, but if it costs £20 per tonne then it must be efficiently utilised. It is not enough either, that one's grassland management is good. Whole farm management must also be of a matching quality when one moves into high farming.

ECONOMICS OF NITROGEN USE

This emphasis on efficiency draws a sharp line between the economics of nitrogen use on dairy farms as opposed to fat lamb and cattle feeding farms. The very much lower efficiency of meat as opposed to milk production does not make it generally possible to follow a high-nitrogen policy on meat producing farms. A fifty per-cent increase in grass production will not, in the first place, be matched by anything like a fifty per cent increase in gross income. In the second place, the limited margins one has to work with necessitate very cheap food, if a reasonable profit is to be made. This means, in effect, an acceptance of a lower rate of stocking as a consequence of relying mainly on clover nitrogen.

This does not mean that there is no place for nitrogen top-dressing on the grassland of the sheep and cattle farmer. Its use is justified on grass crops intended for conservation or for some early bite, when this will replace more expensive alternative foods, and for the production of foggage where this is used in the back-end as a substitute for a more expensive forage crop. The use of nitrogen will, however, be much more opportunist than it will be on the intensively stocked dairy farm.

It has already been pointed out in Chapter II there may be some place for successive spring top-dressing of nitrogen in fat lamb production using the creep system, where there is a very different efficiency of grass use from that obtained with more normal systems of fat lamb production. But, here again, it will be opportunist use, the criterion being the provision of an adequate succession of leafy growth to meet the needs of lambs.

It is stressed that if one is going to rely on clover nitrogen, a very real effort must be made to encourage a vigorous growth of clover in the sward by using all the management devices at a farmer's disposal. We have been thinking in terms of a pasture which will produce 6,000 kg of leafy dry matter without the aid of fertiliser nitrogen, but this is substantially in advance of the yield of the average pasture. The average pasture, however, has insufficient clover because of failure to provide adequate lime, phosphate and potash, because it has been taken too frequently to an over-mature stage before cutting, or because it has not been properly grazed.

The famous fattening pastures of this country invariably have a good balance of grass and clover, because these are on land of high natural fertility and they are always scrupulously managed. Good grazing management can be applied to any pasture, and if the soil is deficient in some essential plant food it is no great problem to make this deficiency good. As the late Dr William Davies said on many occasions referring to grass on lowland farms, there is very little poor land in Britain, but a lot of poor grassland management. Fortunately this situation is changing rapidly as land becomes more expensive and there are other economic pressures operating to encourage a better appreciation of the importance of well-managed grass in improving the viability of livestock farms.

CHAPTER VI

PASTURE PRODUCTIVITY

THERE are two main aspects to productivity, namely the total quantity of dry matter that a pasture produces, and the quality of that dry matter in terms of digestible nutrients. In addition, the herbage must be palatable, and it must be safe in the sense that it does not impair the health of stock or, in the case of dairy cows, produce milk taints.

We do not yet know enough about certain health aspects, and for some farmers the fear of bloat or grass tetany is still a deterrent against intensive grassland farming. Fortunately, trouble of this nature is not continuous and widespread, but every now and then some farmers run into real trouble. However, one can be reasonably hopeful of an early break-through on these problems, with the steady advances that are now being made in rumen physiology.

FEEDING VALUE OF GRASS

Broadly the same factors which influence yield also affect quality, namely botanical composition, soil fertility levels, and management, with this last factor having an over-riding influence on the other two. It is not enough to establish desirable species on a well-fertilised soil. One must always offer herbage which is at a stage of growth where it will do stock most good, and this calls for an understanding of the considerable chemical changes that take place in pasture plants from the first stage of active spring growth through to full maturity.

The work of Woodman at Cambridge and Fagan at Aberystwyth defined these changes and related them to changes in feeding value. As pasture matures, there is a progressive fall in its protein content, which is accompanied by an increase in fibre. There is an accompanying fall in digestibility, though digestibility is remarkably well maintained up to the point of seed head emergence, when a rapid deterioration takes place.

The dry-matter content of herbage that is still in the vegetative phase of growth has all the attributes of a concentrate. It is extremely digestible, and it is rich in protein, often to the point of being in excess of the animal's needs. However, it is cheap protein, and so one need not be concerned that good pasture is slightly unbalanced in this sense. The very high quality of grass dry matter at this stage prompted Woodman to advocate the artificial drying of grass to provide a concentrate which would compete with protein-rich feeds such as linseed or soya-bean meals.

At ear emergence, pasture still retains a considerable amount of leaf and its digestibility is reasonably high. After flowering and the maturing of seed, it quickly moves to a stage where it is little better than straw. Not only is the protein content low, it also has a very poor digestibility because the soluble or easily-digested carbohydrates like sugars and starches have been converted to fibre, with its low-energy availability to the animal. Intake of dry matter is also impaired by this lower digestibility because of the slower passage of food through the alimentary tract.

Work at the Hannah Dairy Research Institute and elsewhere has shown that the deterioration in feeding value in pasture as it ages is due to more than reduced digestibility. Mature herbage, such as hay or silage that has been cut after flowering stage, gives rise to a very different sort of rumen fermentation from that resulting from eating young herbage. With young herbage the fermentation is very similar to that which occurs with concentrated foods. The end-products have a very high efficiency for production purposes, whereas the end-products of digestion of mature herbage, though they are valuable in satisfying maintenance requirements, are very inefficient for production purposes.

If a farmer is able to offer his cows only hay or silage made from very mature herbage, then it is necessary, if he is going to obtain a reasonable level of production, to feed concentrates at a heavy rate. If, on the other hand, he is able to provide well-made silage, cut at the flag stage, and hay that has been cut at the early flowering stage, he will be providing bulk foods with a concentrate-sparing function, because the products of their digestion have much the same attributes as the products of concentrate digestion.

With a knowledge of these facts, it is now possible to define two primary objectives for a grassland farmer. The first is a provision of a succession of leafy herbage over the grazing season. The second is the conservation of surpluses over and above immediate grazing requirements at a stage of growth that will give them, in the form of hay and silage, a production as well as a maintenance function.

Fortunately, the two objectives are fully compatible, and their attainment will secure the maximum production of nutrients over the year. The aim is to use the grazing animal and the mowing machine to keep pastures juvenile and in active growth. If one delays cutting till an advanced state of maturity, aftermath growth will be slow, because young tillers will have been smothered and there will be a time lag before fresh tillers take their place. One should try to avoid leaving cutting to the point where there is a yellow bottom to the sward. The aim should be to retain some greenness, for this will ensure a quicker recovery.

Herein lies one of the great advantages of ensilage over haymaking in this doubtful climate of ours.

Though, in theory, it is right to cut grass for hay when it is still full of nutriment, unfortunately at this stage it is also full of sap and is inclined to pan down in the swathe. This in turn slows the rate of drying. On balance, it is much easier to make good silage than good hay earlier in the season, but at the same time it pays to be an opportunist. If, as so often happens, good haymaking weather comes in late May, it is sensible to turn from silage to haymaking, provided one has equipment that will fluff up the cut grass and speed the drying process.

In this context, however, the type of conservation product is irrelevant. The important issue is a succession of fresh aftermaths being made available for grazing as the season advances. The man who leaves his mowing until July will have a large area of poor aftermath coming in at the one time, and he has lost much of the benefit of using the mowing machine as a tool in pasture management.

This last point is of great consequence where one is using intensive methods of grazing, such as folding behind an electric fence or rotational grazing,with subdivision into small paddocks. After the second grazing, a considerable proportion of the pasture will be fouled by excrements and will be neglected by the stock. No amount of grassland harrowing will correct this. The best approach is to endeavour to alternate mowing with grazing, so that one combines the operation of building up winter reserves with the maintenance of a leafy succession of clean growth which will be attractive to stock.

BOTANICAL COMPOSITION

Turning now to the influence of botanical composition, there appears to be surprisingly little difference in feeding value of the more common pasture species, at corresponding stages of growth. The main feature which determines whether a species is good or bad is duration of leafy growth. A grass like common bent is late in starting spring growth, it moves fairly quickly to the reproductive stage, and it has a limited capacity for autumn growth. A pasture type of ryegrass, on the other hand, has a much longer growing season, and with comparable management has a much more favourable leaf-stem ratio, as well as a much greater productivity.

At the same time, it must be stressed that good management can make a considerable difference to the value of what are generally considered to be inferior grasses. For instance, on peat soils in New Zealand extremely high production is obtained from pastures which contain large proportions of Yorkshire fog. Leafiness is maintained by intensive grazing for short periods, which are followed by longer periods of rest.

We have, unfortunately, little reliable information on the relative feeding values of the main grass species, let alone differences between varieties within species. Some work at the Hannah a number of years ago indicated a much lower dry matter intake by cows grazing cocksfoot as compared with perennial ryegrass. Recent digestibility studies on S 37 cocksfoot and S 23 and S 24 perennial ryegrass at the Grassland Research Station, Hurley, have shown cocksfoot to be at a considerable disadvantage as compared with the ryegrasses, at all stages of growth. Even in April, when over 80 per cent of the herbage consisted of leaf, the digestibility of the cocksfoot was only 70 per cent, as compared with 75 per cent for the S 24. This difference persisted through to the fully-mature stage in June, when there was virtually no growth. Similar differences were maintained in the digestibilities of the monthly regrowth. Both ryegrasses maintained digestibilities in the 70–75 per cent range from June to October, but the cocksfoot figure was never better than 70 per cent, and fell to as low as 60 per cent in October.

After seeing these results, and the suggestion from the Hannah work that cocksfoot is less palatable than ryegrass, it is not surprising that so many intensive grassland farmers have such a poor opinion of cocksfoot for production purposes. Cocksfoot

PLATE 5a

Phosphate deficiency leads to lack of clover in the sward. Grass on right had received no phosphate since sown three years before, while that on left received phosphate at the end of the third year and shows marked clover growth.

PLATE 5b

Professor Martin Jones transformed uniform pastures into diverse types by variations in timing and intensity of grazing and cutting.

PLATE 6

Mechanical damage is most likely to occur on wet soils. It opens the sward and promotes conditions favouring the spread of light-demanding weeds such as buttercups and daisies.

appears to be a better friend of the cake merchant than it is of a farmer who is trying to obtain the maximum amount of milk from grass. That certainly is our impression, from experiences with cocksfoot in both the south and the north of England.

Though Hurley work shows that the digestibility of S 23 is maintained until well in June, when ear emergence takes place some three weeks later than S 24 ryegrass, it is unfortunately much too late in starting growth in the spring. Its palatability is in some doubt as well. In our experience it does not seem so attractive to sheep and cattle as some of the Continental pasture strains of ryegrass, such as *Melle* and *Mommersteeg*. It is not enough to have grasses which have good digestibility values. They must also be attractive to the stock that graze them. Unquestionably, much of the value of the timothy/meadow fescue ley for milk production is attributable to its palatability, and the same is true of Italian ryegrass.

FACTORS AFFECTING PALATABILITY

Species or variety alone does not determine palatability, which is also influenced by stage of growth as well as freedom from any form of fouling. Any quick-growing leafy pasture which is free from objectionable plants like stinking mayweed will invariably have a high degree of palatability. So those conditions which promote active growth, such as adequate moisture and soil fertility, also promote palatability. Palatability with its effect on intake is probably one of the main reasons why a direct reseed in its maiden year has such a capacity to stimulate milk yields.

Variety in a sward also seems to contribute to its palatability. At Wye College a realignment of fencing resulted in the inclusion of a small area of cocksfoot in a mainly ryegrass field, and a small area of ryegrass in the adjacent field sown to cocksfoot. It was remarkable that the hardest grazed parts of both fields were these two small transfer portions.

The habit of stock fossicking along hedgerows probably arises from their liking for variety, though some people attribute this to a craving for something that highly improved pastures lack. Probably this assumption goes too far. One cannot, for instance, conclude that because cows which break into a garden seem to have a penchant for one's most precious plants that these should be included in pasture seeds mixtures.

MINERAL DEFICIENCIES

There is, however, a school of thought that believes that mineral-efficient herbs such as burnet, plantain, chicory and yarrow should be included in ley mixtures or should be sown in strips down the field to minimise the alleged dangers of too much purity. The literature, however, is conspicuously short of convincing arguments on the wisdom of including these herbs, apart from the fact that they are generally richer than grass in such minerals as calcium and magnesium. White clover also shares this quality of being richer in these minerals than grasses, and it has the additional advantage that it will fix atmospheric nitrogen.

A trial at Cockle Park, which compared a clover-dominant sward with a normal ryegrass/white clover sward, both with and without sown herbs, did not reveal any advantage from the inclusion of herbs, as measured by the thrift of grazing ewes and lambs. If anything came out of the early stages of the trial, it was the advantage of having a high proportion of clover in the sward, for in the first year the best lambs came off the very clovery swards.

Generally speaking, there is little to worry about in respect of mineral content of swards which have a good balance of grass and clover and are well fertilised. There are, however, exceptions to this rule, and the most notable relate to those areas which are known to have a deficiency of copper or cobalt. Pasture improvement on such areas seems to aggravate these trace element deficiencies, for the greater growth of herbage dilutes even further the small quantities of available cobalt and copper.

Perhaps the most spectacular example of cobalt deficiency was that of millions of hectares of so-called "bush-sick" country in New Zealand, where cattle and sheep would waste away on what appeared to be first-class pastures. Then it was established that the difference between death and normal health on these pastures was a matter of a few parts of cobalt per million parts of dry matter. Normal rates of top-dressing with superphosphate, with cobalt salts added at the rate of a kg per tonne of superphosphate, amply safeguarded the pastures. A similar correction of copper deficiency has been achieved by the application of "copperised" superphosphate.

We have areas in Britain which are either copper or cobalt deficient, but these are fairly well defined, and symptoms of deficiency are quickly recognised by veterinarians. One of the most likely areas for a copper deficiency is on reclaimed peat—in fact, in some parts of the world, copper deficiency disease is described as "peat scours." Prevention may be effected by drenching or by the provision of mineral licks, but the most convenient and certain method, if deficiencies are known, is to incorporate the missing element in a fertiliser which is regularly applied to the pasture.

GRASS TETANY

One of the most intriguing, and one of the most serious, problems of mineral deficiency associated with the grazing animal is grass tetany, or hypomagnesaemia. A condition of the disease is the very low level of magnesium in the blood. Because of this, it can be loosely classed as a deficiency disease, but not in the same category as cobalt deficiency.

The disease is generally associated with two distinct periods—one in the winter prior to active pasture growth, and the other during the spring when growth is active. There is also a liability of the disease occurring during the autumn, again when grass is making fairly rapid growth. It is scarcely correct to call the disease grass tetany in the late winter when the intake of grass is very low. Then it is probable that the occurrence of the disease is due to a very low intake of magnesium, because of the nature of the diet, and the action of certain stresses such as those caused by the demands of lactation or difficult weather conditions. The other period of grass tetany, namely that during the period of active pasture growth, is of more concern to progressive grassland farmers.

Knowledge of the disease and the conditions that are associated with it is at that annoying stage where nobody can be categoric about it, certainly not to the point of laying down definite instructions for the complete avoidance of trouble. Young, quickly-growing pasture appears to be particularly suspect, and in the view of some farmers this implicates nitrogenous fertilising. Whether this is a correct suspicion or not has not been satisfactorily established. Though the disease occurs in many herds which are grazing early-bite pastures which have received nitrogen, there are also many herds on the same sort of grazing which are free of the trouble. It is certainly not a black and white issue.

Work in Eire and the Netherlands has incriminated high nitrogen usuage when this is associated with heavy dressings of potash. This boils down to the same thing as heavy dressings of nitrogen on soils with large reserves of potash, which may have been created by heavy stocking and the return of urine over a period of years—a condition that is likely to be found in home paddocks.

The biochemistry of the situation is not yet fully understood, but it is known that if there is a high availability of potash during a period of active growth, there is a luxury uptake of this element at the expense of magnesium. This is especially true of grasses as opposed to clovers, and it is grass rather than clover that makes the main contribution to grazing in the early spring when nitrogen has been used to stimulate growth. No-one can say for certain, however, that a low magnesium content of this early spring growth is the direct cause of grass tetany. There may be other factors operating, for example upsets in the animal's normal physiological processes which impair its capacity to mobilise the considerable reserves of magnesium it carries in its body.

Whatever the ultimate explanation, the fact remains that the onset of the disease can be really terrifying. In the spring of 1952, at Wye College farm, we were tremendously proud of the fact that our sixty-cow Ayrshire herd was completely supported by grass in April and was producing 1000 litres per day. This was a very satisfactory level, in view of the fact that the herd consisted mainly of autumn calvers and included a higher than normal proportion of heifers.

Then we struck trouble. Four high-producing cows died suddenly, two of them within minutes of having given their usual amount of milk, and nearly half the herd developed a "jittery" condition which one associates with maltreatment. The veterinarian who attended the herd had no difficulty in diagnosing the trouble, because it was a copy-book case of hypomagnesaemia. Affected cows were injected and the herd was moved immediately from the quick-growing ley on which they had been grazing to an old permanent pasture. In addition they were given a small ration of concentrates fortified with calcined magnesite, so that each cow got at least 57 grams of this mineral per day. The condition disappeared almost as quickly as it arrived, though the herd did not return to its previous level of production.

Which of the remedial actions was the most efficacious no-one can say, but the whole incident left one indelible impression, namely the importance of making gradual changes in the feeding of livestock. One cannot hear a farmer say that he has suddenly dropped supplementary feeding in the early spring because he has ample grass without giving him the warning of this experience. This does not mean that we no longer advocate the provision of early bite through the growing of a species like Italian ryegrass, which receives a heavy dressing of nitrogen. To advise another course would be as rational as an advocacy of a return to the horse and cart to cut down the incidence of road accidents. The tempo of our farming is such that we have to use such aids to lower the cost of milk production, but there are certain limits to be observed until such time as we have a better understanding of the complexities of ruminant nutrition.

Until we are in a position to produce really safe early bite, it is wise to continue to feed some additional food which has been fortified with calcined magnesite until the grass has "hardened," as it does in May. Significantly, the safe period for grass coincides with the time when clover starts to make an appreciable contribution to grazing.

Clover, however, is not only possessed of virtues, for dominantly-clover swards are seriously implicated with the occurrence of bloat or hoven. Fortunately, in Britain, bloat is not the nightmare it is in certain parts of New Zealand, and under our conditions of management it can hardly be described as a really serious hazard over the general run of farms.

AUTUMN GRASS

Any account of pasture quality is incomplete without some discussion of differences between spring and autumn pasture. Chemically these are small, though sugar levels may be slightly higher in spring pasture, and yet there is a considerable difference in feeding value in favour of spring pasture. Probably the main reason for the difference in quality is the rather higher moisture content of autumn pasture, which is aggravated by the presence of a considerable amount of free moisture as the result of heavy dews. Comparison between dried grass from the two seasons, cut at corresponding stages of growth, reveal practically no difference in feed value.

Under grazing conditions there are probably other contributory reasons. There is a greater liability of autumn grass being fouled and, in consequence, having a lower palatability. There are also greater risks from such infections as husk or intestinal worms, which appear to be more potent in their effects in the autumn than they are earlier in the season.

There is, of course, little that the farmer can do to improve the quality of autumn grass to the point where it is comparable with spring grass. He has to accept its relative inferiority and make allowance for this in his supplementary feeding plans, especially when he is dealing with freshly-calved cows—a point which will be discussed in greater detail at a later juncture. At the same time it has to be stressed that autumn grass is still capable of making a most valuable contribution to the feeding of a herd or flock.

It is well worthwhile to make a special effort to build up reserves of back-end grazing and to ration them so that they are not wasted. When the last of the autumn good grass is finished, even though it may be as late as December, one invariably sees this reflected by a drop in milk yields, despite efforts to make compensating changes by the provision of other food. It may well be that autumn grass is equally valuable to some farmers as early bite, but this is a point we will argue about later in the book.

CHAPTER VII

PASTURE UTILISATION

EFFICIENT use of pastures depends on seven principal factors:

(1) Inherent productivity of the livestock that it carries;
(2) Quality of management of this stock;
(3) Health status;
(4) System of grazing that is adopted;
(5) Efficiency of conservation practice;
(6) Levels of feeding supplementary to pasture;
(7) Intensity of stocking.

The first three factors are outside the scope of this book except insofar as health considerations and general thrift are affected by management decisions. For instance, in the rearing of young stock—be they calves or lambs—worm burdens can be materially reduced by appropriate resting of pasture to break the cycle of infection. Again, both fertiliser practices and the feeding of supplements in the spring can affect the level of incidence of hypomagnesaemia. These are matters of great concern to a farmer who is trying to get the best out of his grass.

We shall be dealing with two further factors, supplementary

feeding and conservation practices, later in this book but they are mentioned here principally to stress the fact that none of these factors can be considered in isolation from the rest because they interact together. The economic utilisation of grassland is a complex where the successful farmer combines his skill of stockmanship with sound management decisions relating to the handling of both stock and pastures.

The remaining two factors, intensity of stocking and system of grazing, are very closely linked and it can generally be accepted that the higher the intensity of stocking, the more important it is to have controlled grazing, whether this is achieved by folding, using an electric fence or more permanent subdivisions to permit some form of paddock grazing. The more stock one has on a farm, the more important it becomes to have a planned system of pasture production and a means of rationing the available grazing to minimise wastage in its various forms.

GRAZING MANAGEMENT SYSTEMS

We can distinguish these forms of grazing management:

(1) Set stocking where a herd or a flock remains on the same field over a long period, sometimes for the greater part of the grazing season.

(2) Rotational or paddock grazing where there is close subdivision and stock are concentrated in one enclosure for short periods before being moved to a fresh enclosure.

(3) Fold grazing where stock are given a ration of grass, usually at least once daily behind an electrified wire which can be quickly moved to its next position.

There are variants of these systems. Rotational grazing can be integrated with conservation on the same area or, in dairy farming, a two-sward system can be practised using one area which is principally used for conservation. Later we will be discussing the pros and cons of these different approaches. Another variant of rotational grazing is that developed for intensive fat lamb production, known as forward creep grazing, where suckling lambs have preferential access over their mothers to the next paddock in the rotation.

Set stocking is generally characteristic of less intensive forms of utilisation, e.g. the grass fattening of bullocks where equation of appetite and availability of nutrients is achieved by adjustments of stock numbers. It is generally recognised that store cattle finish more quickly if they are subjected to the minimum of disturbance and this is usually the situation with set stocking. The same applies in fat lamb production but inevitably, though individual performance is better, there is a loss of production per unit area because it is not possible to maintain a really high intensity of stocking.

Stocking intensity is unquestionably one of the key factors in determining relative efficiency of pasture utilisation and it can be taken as axiomatic that maximum production per ha is not coincident with maximum output per individual. In other words, in order to effect the highest level of conversion of grass nutrients into milk or meat or wool, there must be some sacrifice of individual performance but this must never proceed to the point where net returns are impaired. The intensive grassland farmer has to walk a tight rope in this connection, and especially is this true in dairying, but fortunately there are aids in the form of supplementary feeding to help him in his balancing act.

ROTATIONAL GRAZING

Almost the best approach to an understanding of this problem comes from an examination of the advances that have been made in intensifying pasture production that have occurred over the past sixty years and a key issue in this has been the development of rotational grazing.

Some years ago, Sir Bruce Levy, when addressing the Farmers' Club, stated that intensive rotational grazing gave fifty per cent more production than uncontrolled stocking of grassland.

His statement was not questioned by those present, though he had little direct experimental evidence to support it. His views were based on many years of experience over a critical period in the development of grassland farming in New Zealand. As a young man, he saw the pioneering phase of the industry, when farmers were grazing their stock extensively amid blackened stumps on pastures that depended largely on residual fertility of the bush burns. But by the time he made this statement New Zealand had achieved the status of being one of the foremost grassland countries.

In the interim, the German development of the Hohenheim system of intensive rotational grazing, and Woodman's work on intervals of cutting and its influence on productivity, had a tremendous impact on New Zealand grassland practice, especially on dairy farms. The pattern of development was one of closer subdivision of farms so that there were as many as twenty paddocks, each of which was grazed in turn over a period of 1–5 days, according to season, first by the milking herd and then by followers, and then rested for a period of 2–3 weeks. The aim was to graze at the leafy 100–150 mm stage, and to conserve the spring surplus of silage or hay by dropping a proportion of the paddocks from the grazing rotation.

It was a beautifully simple system to advocate and it was, moreover, highly successful. Top farmers on the better land were carrying a dairy cow and her share of replacement stock, and producing 7000 litres of Jersey milk on a ha of grassland without the aid of supplementary crops or concentrates. The plough, in fact, became a rarity on these farms, and the whole of the feeding was based on permanent grass.

There were many factors other than the system of grazing contributing to this spectacular improvement, for example, the effects of herd recording, the improvement of stock, better management and control of disease, and a greater use of phosphatic fertilisers. But undoubtedly the system of grazing played a substantial part.

However, it is highly doubtful whether the direct effects of quick grazing followed by rest were as great as Woodman's findings would have led one to expect in determining this improvement. Cutting with a mowing machine, especially when the sward is bared very closely (as it is with the lawn mower commonly used in small experiments), is a very different matter from defoliation by grazing, especially when the animals concerned are cattle. Not only is there more leaf left to effect recovery, but the pasture under continuous grazing tends to assume a more prostrate form, which, though short, can be remarkably leafy—a condition seen clearly on farms adopting intensive set stocking with sheep. In addition, the mowing machine does not return excrements, with their stimulating effects on growth.

Weight is added to this view by results obtained by Australian workers who have established, under their conditions, that continuous grazing, provided it retains a reasonable leaf area, does not suffer in comparison with rotational grazing in respect of nutrients produced from a pasture. There is also confirmatory evidence from a long-term New Zealand trial at Ruakura with dairy cattle, which will be referred to in more detail later, that there is no appreciable difference in dry matter production from pastures subjected to either continuous or controlled rotational grazing, provided there is full utilisation without over-grazing.

PLANNING OF FOOD SUPPLIES

Undoubtedly the main advantage of controlled grazing is derived from two principal features of the system. The first is what one may call a rationalisation of the farming programme to ensure a continuous supply of good food over a long grazing season, with the opportunity of conserving surplus grass for those periods when growth is inadequate for the needs of the herd. The second is the confidence that such a system gives one to increase the intensity of stocking, which, in our opinion, is by far the most important single factor in determining degree of pasture utilisation.

These points can be clarified by a consideration of a situation where a herd or flock is continuously grazing one large field from early spring to late autumn. At the point of commencement of growth there will be inadequate feed, and each blade of grass will be pruned as it comes within grazing range. Of necessity, the stock have to receive supplementary feed to keep up production.

By mid-spring, there is an equation of stock appetite and grass growth. But within a matter of a few weeks, as grass gets ahead of stock, the field is a mosaic of closely-grazed and neglected patches, with the latter running up to head to deteriorate in quality. There is a surplus in the field, but it is not practicable to cut it for silage, and the best one can do is to top it, or leave it to provide a bit of rough grazing if there is a summer drought. In this event it will have a limited value in maintaining stock, but no worthwhile productive function.

More often than not, it will remain till late autumn, when it may not be cleared by grazing. In the meantime, it will have affected sward composition, for less desirable grasses for production, like Yorkshire fog, *Poa trivialis,* and cocksfoot, will have achieved dominance in these rough patches and clover will have been suppressed. Too often, these rough patches are carried over into the spring, aggravating the situation and accelerating the process of sward deterioration.

Contrast the management of that field with a subdivision either by permanent or temporary fencing. Each area can be grazed in turn, taking precautions to avoid over-grazing, until there is a surplus of grass. Then a proportion of the paddocks can be withdrawn from the grazing rotation for mowing. As these come back into grazing with clean fresh aftermaths, paddocks which have been grazed twice can, in their turn, be closed for conservation and cleaning, which is achieved by the one act of cutting. The summer period of reduced growth brings every paddock back into the rotation. If it is necessary at this time, as it is with breeding ewes, to keep the condition of stock down, they can be enclosed in the paddocks with rough feed to turn this into dung and urine, which will mobilise fertility to promote fresh growth, and correct any tendency towards imbalance of species. The autumn flush can then be systematically rationed with the minimum of wastage by trampling and fouling.

One sees in this pasture the rationalising of feeding that has already been referred to. First of all, there is the safeguarding of quality, because high-producing stock, within the limits imposed by season, are always offered grass at the optimum stage of growth nutritionally. Secondly, there is a high proportion of surplus nutrients which are conserved, almost as a by-product of pasture control. Thirdly, there is an opportunity with these small independent grazing units of ringing the changes by management, top-dressing, and the use of special-purpose seeds mixtures, if these are considered important, to give a high measure of flexibility.

Essentially this was the basis of the system of intensive paddock grazing that the late Andre Voisin demonstrated in Normandy some years ago and which now has many advocates in Britain. An important feature of the system is that it allows preferential treatment of stock that require the best available grazing. In its earliest form in New Zealand, the milking herd took the cream off the pasture, and dry stock following behind grazed the residue. Later a scheme was developed where the calves were grazed ahead of their mothers to give them an advantage of approximately 50 kg at the yearling stage over calves which were reared on the traditional system of a calf paddock which denied them the opportunity for reasonable development. This is a point to which we will return later.

THE RUAKURA EXPERIMENT

The Ruakura experiment on controlled versus uncontrolled grazing, already referred to, is one of the most thought-provoking studies in pasture management that has been attempted. Originally it was designed largely as a study of the influence of plane of nutrition on dairy cows maintained entirely on pasture. The pasturage was divided uniformly into equal areas to provide two farms. One farm system was based on controlled rotational grazing with subdivision, the conservation of surpluses, and the saving of autumn feed for late-winter grazing of newly-calved cows. At the outset of the experiment, it represented what most people considered to be a climax in grassland dairy farming.

Its partner farm had the same rate of stocking, the same sort of stock, and the same husbandry, except that there were only two fields, a night and a day enclosure with a portion of the latter cut off in the spring to make hay, which was the only form of supplementary winter feeding. This represented a slap-happy system of dairying, insofar as pasture management was concerned, but it is important to remember that all other aspects of management were good.

At the end of ten years the surprising result was obtained that production was only 13 per cent higher on the controlled system than it was on the uncontrolled. It is not surprising that the result made those concerned wonder about the economics of a system based on close subdivision and intensive grazing. Detailed examination of the results and the conditions of the trial, however, reveal several important features which do not weaken, but rather strengthen, the case for controlled grazing, especially under British conditions.

The first point is that, though in good grass years there was very little difference between the systems, because there was no real stress on the cows, in bad grass seasons, with a delayed spring and summer drought, the difference widened to 25 per cent. A good grass year in New Zealand means good grazing for virtually ten months of the year and very little nutritional stress on stock, even under extensive grazing conditions. A bad grass year is something much more akin to the conditions we usually have to put up with, especially in regard to length of winter.

IMPORTANCE OF STOCKING INTENSITY

Controlled grazing has a valuable insurance function in difficult years, because it is a method of preventing wide fluctuations in annual income. More than this, it gives confidence in building up the stocking rate on a farm. If one is coming out of the average winter with substantial reserves of hay and silage, the fear of over-stocking disappears.

This leads to the second feature of the trial. Both farms carried the same number of cows and followers at the rate of one cow-equivalent per 0.4 ha. Though this was high, even on the existing New Zealand standards, on the controlled system it was still within the range where production did not suffer because of over-stocking. The point of this remark will be understood when it is stated that subsequent adjustments to the experiment, involving increases in stocking rates (but the elimination of followers), resulted in a reduction in yield per ha on the uncontrolled farm, because the fall in yield per cow was not compensated by the extra stock which were carried. This did not happen on the controlled farm, for, despite a small fall in production per cow, production per hectare rose to the remarkable figure of 12,716 litres of Jersey milk to the ha.

Another feature which helped to account for the smallness of the difference between the systems in the first ten years of the trial was that the cows on the controlled system were approximately 50 kg heavier than those on the uncontrolled. This extra weight used food that would otherwise have gone into milk, and it raised the maintenance requirements of the controlled herd. Here we see not only the importance of small animals under grazing conditions, but also evidence of the operation of diminishing returns in the feeding of dairy stock. As the plane of nutrition is raised, the proportion of nutrients that goes into milk tends to fall because more goes on the cow's back. Although these cows look well, and produce well, the extra production is not as great as it would be if there was a straight-line relationship between food input and milk output.

PLATE 7a

Valuable spring grass often has to be rationed by sub-division with electric fences. These cows were grazing RvP Italian ryegrass at Frondeg Farm, Welsh Agricultural College, on April 5th, 1973.

PLATE 7b

Another view of the Welsh Agricultural College's spring-calving herd on RvP Italian ryegrass in the first week of April.

PLATE 8

Yearling spring-born steers turned out to grass on the Welsh Agricultural College's Tanygraig Farm, March 1973.

There is evidence from at least five other major New Zealand experiments that a high rate of stocking is more important than a high yield per cow in determining high production per hectare and this is borne out by the results of farm surveys. This does not imply that yield per cow is unimportant, nor that there are real dangers of over-stocking. Every effort must be made to improve cow yields by breeding and selection, disease control, and good husbandry, while nutritional stresses on the herd at critical times, such as immediately prior to calving and during peak lactation, must be minimised.

This brings us to what might be called the psychological effects of increasing stocking rate. Once a farmer is committed to a heavier stocking rate, he will make every effort he can to prevent waste and to increase pasture production by the several methods at his disposal so that his stock do not suffer.

The importance of high stocking rates is just as great in Britain as it is in New Zealand, and information is accumulating to support this view. It is a reasonable goal on a British dairy farm, with land of reasonable quality, to aim at a cow equivalent to 0.5 ha of grassland with concentrates, either purchased or home-grown, being fed at a rate of not more than 750 kg per animal, unless the herd and its management are of exceptional quality.

That relatively few achieve this sort of standard is attributable partly to poor pasture management, but also to the arbitrary ceiling that is put on farm carrying capacity by the accommodation available in buildings. If a 40 hectare grass farm has standings for thirty cows, and yards and boxes for a similar number of young stock, then there will be approximately a ha to the cow-equivalent, and the tempo of the grassland farming will be geared to this ratio.

One of the great advantages of the loose-housing system of cow-keeping, especially with the rather extravagant area-per-cow allowances that have been advocated by the Ministry of Agriculture, is the flexibility it gives to increase cow numbers, because one is not tied to a number corresponding to the availability of standings.

CASE FOR CONTROLLED STOCKING

It is stressed, however, that if one attempts a high rate of stocking in Britain, then it is absolutely essential to have controlled grazing—either by permanent sub-division, which is appropriate on the all-grass farm, or by the use of electric fencing on the larger fields that are characteristic of alternate husbandry.

In this connection it may be said quite categorically that fold grazing, using the electric fence, especially if the precaution is not taken to back-fence, does not give any appreciable advantage over intensive rotational grazing, using paddocks.

The early experiments contrasting fold grazing with rotational grazing in this country indicated a 20–30 per cent production advantage in favour of the former system, but this difference was entirely accounted for by the increased stocking rates which were adopted for fold grazing. New Zealand experiments of a similar type took the precaution of equalising stocking rates, and there were no appreciable differences in milk yield per ha. It is important to remember that the comparison was made with *intensive* rotational grazing and not *extensive* rotational grazing, where the herd is in the one field for a number of days. Here fold grazing is likely to be a very much better proposition.

However, it serves little purpose to debate the merits of the two systems, which will in any case vary according to the circumstances of the farm. The all-important consideration is that there is a control of grazing to create a situation where a farmer can see weeks ahead in his grazing programme and be able to plan so that there is a sequence of quality grass always available for his herd. At the same time, he must be in a position to isolate surplus grass from his stock and conserve it so that winter carrying capacity is no longer a bottle-neck limiting summer stocking to the point that there is serious under-utilisation.

TWO-SWARD SYSTEM OF MANAGEMENT

This system has acquired a lot of prominence on dairy farms

over recent years, largely because of its advocacy by those who are interested in promoting nitrogen sales. Basically it represents an attempt to use one closely subdivided part of the farm for grazing and a second part for conservation, usually in the form of two silage cuts, with growth during the remainder of the season being grazed, generally by followers. It is by no means a new concept. Somerville when he laid down the Palace Leas manuring trial at Cockle Park in 1896 was doing no more than attempting to find the best method of increasing yields from fields that were used for hay year after year.

The perpetual hay-field fell into disrepute for a very good reason because repeated depletion of plant nutrients, without an adequate return of fertility, resulted in miserable crops that contained more herbs and weed grasses than species that are generally recognised as being useful. Now the wheel has taken a full turn, as it so often does in agriculture, and we are back to a two-sward system in a different guise. Fortunately it includes recommendations for adequate fertilising of pastures and this is important on those fields where there is an incomplete return of animal excrements.

The proponents of the two-sward system extol its simplicity. The grazing area is divided into 20–30 enclosures, the number varying according to taste and convictions as to how quickly the cows should return to the paddocks. Usually the herd occupies a paddock for a day and a night, and any grazing that is left behind is regarded as something for the next time until mid-season when there is generally a need to do some topping. Usually it is recommended that nitrogen is applied after each complete grazing so that there will be a good growth for the next visit by the herd and this largely accounts for the support the system gets from those in the nitrogen business.

The pro-arguments include the fact that there is no need to shift an electric fold fence either once or twice daily, and it is also claimed that there is less need for a dairyman to exercise his judgement as to how much or how little he will allow his herd every time he shifts this fence. It is also argued that there is a well-defined conservation area so that the whole utilisation programme, both grazing and conservation, can be planned at the beginning of the season.

There may be much to be said for the system for the not-so-good grassland farmer who still has a lot to learn about how he handles his pastures and his herd. It may appeal to the good farmer, who through pressure of other business, had to delegate responsibility of day-to-day management to someone else in whom he has not complete confidence, but it cannot be regarded as a top level approach to grassland dairy farming.

For our part we would be less critical about this system if there was much less variability in pasture, both quantitatively and qualitatively, over the season. Highly rigid systems of utilisation are not compatible with the dynamics of pasture growth and we prefer a more flexible approach where paddocks are sufficiently large that it is practicable to take conservation equipment into them when this is necessary. Such a size will necessitate the use of the fold fence at the height of the growing season in order to avoid waste of nutrients and also to give more scope for taking paddocks out of the grazing rotation for conservation purposes.

Nor do we favour the 'cookery book' concept of a defined conservation area. Apart from the fact that cutting, especially for silage, is a means of preserving a sequence of clean aftermaths over the summer, a point that has previously been stressed, there is the need to make decisions that match the vagaries of climate. A late spring will often necessitate a raid on an area that may have been earmarked for conservation or, conversely, a favourable spring may give an opportunity for building up fodder reserves.

The higher the stocking intensity, the more important it is to come out of the winter of an average year with reserves of fodder so that there is something in hand for the occasional year when Nature is being difficult. In other words, one has to use one's head in top level grassland farming for there is no one recipe that fits the variety of circumstances that are encountered, especially when there are very high stocking intensities. There is really no satisfactory substitute for brains and a preparedness to use them to best advantage.

GRAZING STORE CATTLE

The evidence on the different systems of stocking with store cattle, which are either to be reared as dairy replacements or are being fattened, is much more meagre than it is for milking cows or fat lambs.

Reference has already been made to the New Zealand system of rotating young calves on pastures ahead of the dairy herd. In the Ruakura trials Jersey calves reared in this way were nearly 50 kg heavier at nine months than their mates reared on the old method of set stocking in a calf field, where there was a deterioration of feed supplies because of highly selective grazing and a build-up of disease. The advantage was not purely one of weight, for the survival rate of rotationally-grazed calves was also much higher.

The early rearing stage in any class of stock is always important, and youngsters, whether they be lambs or calves, should never be used as tools of pasture management. It is a different matter with an in-calf heifer which is well grown. We don't want her to be too fat, and she can be made to work for her living without detriment to her subsequent performance, provided she is steamed up over the last six weeks of pregnancy.

The traditional feeder of beef cattle prefers a system of set stocking to all others, and attempts to equate grass growth and stock appetite by drafting in additional beasts until growth reaches a peak. Then he commences to lighten the load by drafting off beasts in the summer as they become prime. Occasionally a set number of cattle are put into a field in the spring and the adjustment is made by adding ewes and lambs in limited numbers so that the suitability of the pasture for cattle is not impaired.

There have been various attempts to strip-graze fattening cattle, but these have not been very successful. Whether this has been the fault of the system, or the fault of the people operating it, is not clear. The most general criticism, and it seems to be a valid one, is that feeding cattle are restless behind an electric fence, and restlessness is not compatible with a good liveweight gain.

Strip-grazing is of value in a beef unit when handling foggage in the early winter for single-suckling cows. Here the aim is to maintain rather than fatten, and strip-grazing will not only avoid waste from trampling, but will ensure a more even plane of nutrition during the period of utilisation. Rather than give a fresh break every day as in dairying, larger blocks are given, with the fence being moved every 4–5 days.

FAT LAMB PRODUCTION

Again, in fat lamb production the traditional method is non-rotational grazing at low rates of stocking, using cattle to control pasture, and it undoubtedly gives the best lambs. This was the finding in another comprehensive trial at Ruakura which compared non-rotational grazing with rotational grazing at medium and heavy rates of stocking. The best lambs were drafted off the set-stocked system at the lower rate of stocking, but the greatest liveweight gains per ha were produced by the rotational high-rate treatment. Rotational grazing at the lower rate give disappointing lambs, partly because the pastures were not controlled in such a way as to give the dense leafy pasture which is the best sort of food for ewes and lambs. The general conclusion at the end of the trial was that a system of set stocking was preferable where there was a relatively low stock density, but it was advisable for the farmer who was aiming to intensify fat lamb production to adopt rotational grazing.

Work at Wye College with hoggets gave similar results to those obtained at Ruakura. Here the rotation was based on a change of pasture every four days, and it was very noticeable that the sheep spent a lot more time on their feet searching for grass on their last day in a paddock than they did immediately after the change. This behaviour indicates that these were on a saw-tooth plane of nutrition rather than an even plane, and this no doubt contributed to the slightly poorer finish.

These observations contributed to the evolution of forward creep grazing, which is a system where the lambs are always on a high plane of nutrition. It is true that the ewes are on a fluctuating plane, but they are not made to work hard for their living until about the 8th week of lactation, when their efficiency as milk producers had fallen.

One period of the year when rotational grazing of ewes is unquestionably advantageous is during the late autumn and winter, after mating has been completed. Usually at this period there are stubbles which can be eaten out in succession before they are ploughed, and then there are the fogs which remain from cattle grazing. If the ewe flock is concentrated on each pasture in turn, the cream of the grazing can be taken; anything that has been fouled can be left a few weeks to freshen and provide another useful bite for the flock.

An additional advantage of such a system is that it gives a basis for planning a grazing programme, which is not possible where the ewes are spread thinly over all the grassland. Above all, it gives the opportunity of saving some reasonably good grazing, preferably on an old pasture, for the ewes as they come up to lambing. One can work ewes hard during the first two-thirds of pregnancy, but not over the last third, when foetal growth is at a maximum.

This is a typical example of the art of good stock husbandry, stressed at the beginning of this chapter as being essential if we are to make the best possible use of grass.

CHAPTER VIII

FOOD SUPPLIES AND STOCK APPETITE

IN Britain the characteristic pattern of pasture growth, after a long period of winter dormancy, is a rapid increase in spring as soil temperatures move appreciably above 5 degrees C, which is the point at which plants start to grow. The spring flush is generally followed by a summer slump, which is partly attributable to the maturity of plants and, in most seasons, to a shortage of moisture. Favourable conditions in the autumn generally result in a second, but smaller, flush which tails away as cooler temperatures exert their effects. Except in the most favourable localities, there will be no appreciable growth from November until well into March.

The cyclical nature of pasture growth creates many problems for the grassland farmer. But it can be said as a general principle that his aim should be to try and obtain the maximum degree of utilisation as grazing. Any form of conservation has at least two costs. The first is that for labour and machinery, and the second derives from the considerable loss of nutrients involved in both haymaking and ensilage, even when the operations are undertaken efficiently. The average loss of nutrients in silage-making is certainly not less than 25 per cent, and it can be even higher in haymaking by the time the product is fed to stock.

EXTENSION OF GRAZING SEASON

It is important to remember that endeavours to extend the grazing season must be kept in an economic perspective. It is not impossible to grow pineapples on Snowdon, but it is not very sensible to do so, and certainly we don't want out-of-season grass to come into this category of effort. Neither do we want to punish pastures during very wet weather and impair their subsequent productivity in rash attempts to extend the grazing season. There is an old saying in the Midlands, that a bullock out on pasture during the winter months has five mouths.

The fat lamb producer obtains the best synchronisation of grass growth and stock appetite—if we except those fortunate people who can afford to purchase store cattle in the spring and sell them fat as feed fails. The ordinary run of farmers, and especially dairy farmers, are not in this position, and it is necessary that they arrive at the cheapest possible feeding programme consistent with good yields. With present price relationships, this boils down to ensuring that grass in its several forms makes the major contribution to feeding over twelve months of the year.

The principal methods of achieving this end are as follows:

(a) Selective top-dressing
(b) Use of special-purpose pastures
(c) Management
(d) Conservation of surplus grass
(e) Timing of lambing and calving

USE OF NITROGEN

Selective top-dressing implies the application of nitrogenous fertilisers but, it is reiterated, these should be used only when other deficiencies have been made good. The application of nitrogen provides one of the most effective ways of extending the grazing season. The late-winter dressing, applied towards the end of February, in the south, and about a month later in the north, will give effective grazing 2–3 weeks earlier than normal. This comes about because available nitrogen is generally very low at the end of the winter, partly as a result of leaching of this nutrient, and partly because of the cessation of activity by soil micro-organisms concerned in the nitrogen cycle during the period of very low temperatures. Usually it is lack of nitrogen rather than temperature or moisture which limits growth in the earlier part of the spring.

Nitrogen treatments can be used throughout the growing season to keep up the continuity of grazing should this be necessary. Here one comes up against the conflict between bag nitrogen and clover nitrogen, and so it is essential to exercise a considerable amount of judgment. If clover is making a vigorous contribution, and there is ample grazing for forseeable needs, there is no point in using bag nitrogen to stimulate growth.

Except for the growing body of dairy farmers who are practically ignoring the clover contribution and using very large quantities of nitrogen, it is best to adopt an opportunist attitude to this fertiliser during the main part of the growing season, in other words, to use it if it appears that the crop needs stimulation for either grazing or conservation.

Nitrogen has a much more definite function during late summer, because it provides a very effective means of stimulating the autumn flush to provide a bulk of feed which can be rationed well into the winter. Clean autumn grass in quantity will provide sufficient nutrients for the stale cow that has been in milk since the spring. But one can over-estimate its value for the freshly-calved cow, which should receive at least 1 kg of concentrates for every 5 litres of milk it is producing. This need not be a balanced dairy ration, because autumn grass of good quality is high in protein and so the nutritive ratio of the concentrate mixture can be a fairly wide one.

ROLE OF ITALIAN RYEGRASS

The use of nitrogen for stimulation of out-of-season growth is most effective when it is used in conjunction with Italian ryegrass, which has such a good growth performance both early and late in the year, as well as a capacity to remain green fairly well into the winter. One of the most striking developments in pasture farming over the past twenty-five years is the new role of Italian ryegrass. Its traditional use has been as a companion for red clover in a one-year seeds mixture intended for hay production, but now it is very widely used for grazing and usually without the addition of clover.

There is nothing to match it for earliness and its capacity to respond to nitrogen, while it is undoubtedly one of the best grasses we have for stimulating milk yield. On several occasions, both at Cockle Park and Nafferton, we have moved the dairy herd in late May from stemmy Italian ryegrass, which we considered would be better conserved, on to a leafy general-purpose pasture and we have had a drop in production.

The usefulness of Italian for early bite varies, to some extent, according to its age and time of sowing. Undoubtedly the earliest and best grazing in the spring comes from a stand that has been established towards the end of the previous summer that has been properly established before the end of the growing season. Usually such a stand is full of vigour in the following spring, but it suffers from the disability that it will easily poach if the land is heavy and the spring is a wet one.

For this reason, it is wise to have a second stand of Italian in its second year on consolidated land, which can provide alternative early bite should the season be difficult. If one sows RvP, in preference to one of the cheaper commercial strains of Italian, and manages the sward properly, it will thoroughly justify its being kept into a second year in this manner. Usually, after the spring grazing, it will respond to another nitrogen dressing and give a useful cut of silage. It can then be broken for kale or rape or, alternatively, if one prefers to do without these crops, for another reseed of Italian, or possibly a long-term ley.

An alternative sowing time for Italian is in the early spring, either as a direct reseed or in a cereal crop. We adopted this latter method in the cold north-east of England because we found it gave a great deal more elbow-room in planning grazing over the winter-early spring period. The direct reseed in the spring, whether it be Italian or a longer ley, gives a very valuable contribution in the summer, when production is tending to slump, but it gives nothing in April, which is usually a critical month for grass.

Under-sown Italian, which is top-dressed as soon as the straw is removed, will give a very useful stubble feed to take a burden off longer leys in the autumn. It will give further grazing in the spring, at a time when it is advisable to rest the longer leys and so avoid the stresses of hard grazing which are inevitable when a farm is short of spring grass. Such an Italian ley is very cheaply established. No more than 25 kg per ha of seed need be sown, and it can be punished during wet weather without the qualms one has with an expensive ley with a long life ahead of it.

RATES OF NITROGEN APPLICATION

There is no great conformity in the rates of nitrogen dressing used for early bite. Usually they are in the range of 500–800 kg/ha, though occasionally dressings as high as 1,100 kg are given. Rather better results will be obtained by splitting the dressing, using 650 kg for the first application and following it with a further 450 kg after the first grazing. There is always a risk in giving a massive early-bite dressing, in that cold wet conditions may lead to heavy leaching of the soluble nitrogen and much of its value will be lost.

This raises the question of how early one can safely apply this first dressing. Except under the most favourable conditions, there appears to be no point in top-dressing before the end of the second week in February in the south, and 2–4 weeks later in the north. Because we have such variations in the onset of favourable growth conditions from year to year, considerable judgment has to be exercised. Nor is there any point in top-dressing all pastures intended for early bite at the one time. Not only can risks be spread by making a succession of dressings, but it is possible in this way to provide for a succession of grazing.

Usually the nearer we come to the point of ideal conditions for growth, the greater will be the growth response per unit of applied nitrogen. At the same time it is often a worthwhile risk to chance one's arm, so to speak, by top-dressing a small area, perhaps enough for 15 days' grazing, the same number of days before one normally considers it prudent to make the main early-bite dressing.

EARLY GRAZING FROM LONG LEYS

Though Italian ryegrass gives the earliest grazing, this does not mean that it is impracticable to go for early bite from longer leys, especially those based on perennial ryegrass or timothy and meadow fescue. They are particularly responsive if they have been fairly recently established. Almost any seeds mixture which has been established in the previous August will give excellent early bite in the following spring if it has received the appropriate nitrogen dressings. The main drawback against using such leys for early grazing is the risk of excessive poaching.

The growth performance of any pasture in the early spring is very much influenced by the treatment it receives in the previous autumn and winter. Hard continuous grazing at that time, especially if there is mild weather which keeps growth going, will impair spring production.

This brings us back to the point that, so far as the plant is concerned, the function of leaf is not to feed stock but to build up root reserves. A systematic closing up of the pastures in the autumn to produce a lot of leaf will serve two purposes. First, it will give a succession of grazing going into the winter, and secondly, it will build up root reserves to promote early spring growth.

This latter objective will not be secured, however, if back-end grazing is continuous. The aim should be to effect quick utilisation followed by rest, so as to avoid grazing the regrowth thrown up immediately after grazing. Such regrowth, however, can be safely grazed once hard weather sets in and the pasture becomes dormant. In fact, a hard grazing at this time with sheep may be the right treatment for the ley that carries a fair amount of roughness in order to give a clean fresh start in the spring.

Spring growth can not only be impaired by continuous grazing in the previous autumn, but it can also suffer if the pasture goes into the winter in too proud a condition. This is especially true of Italian ryegrass, which can go to mush if it is too soft and leafy when the hard frosts commence. This effect is quite commonly seen in the spring in Italian ryegrass which has been fairly laxly grazed in the previous autumn. Usually the urine patches which have gone into the winter in a very proud condition will be bare of grass which has rotted away. One can see in this danger the importance of adopting a happy medium, by always grazing a pasture in the autumn before it gets dangerously proud, but at the same time avoiding the risk of over-grazing.

The building of the root reserves can be critical at other times of the year, and especially before the onset of the summer drought. Because hard grazing impairs rooting systems, it is wise to top-dress at least one field which has been cut for early silage as a reserve in case there is a June-July drought. If the drought does not materialise it can go for hay or a second cut of silage, but if drought conditions ensue not only does it provide a useful bulk of grazing but, because it has been allowed to build up leaf, it will show a good recovery after this grazing.

Leys based on cocksfoot and Montgomery clover are especially valuable for this purpose on land that tends to dry out in the summer, and the same function can be served by a timothy/meadow fescue ley without cocksfoot on soils with better moisture-retention qualities. Perennial ryegrass is not very useful for this purpose because of its tendency to throw up a second crop of heads.

SPECIAL-PURPOSE LEYS

A few years ago it was fashionable to advocate the use of a wide variety of special-purpose pastures as part of the programme for providing leafy grazing over the season. The wisdom of this is in doubt, because it tends to make management too complicated, and it has the further disadvantage of giving too much in the way of bits and pieces when it comes to conservation, especially on the small farm. A silage clamp can easily become an American sandwich of different sorts of pasture, and this has many drawbacks, especially if self-feeding is adopted.

As a general policy, it is better to rely mainly on one sort of pasture, which can be based on any mixture considered to be suitable for the farm. It can be a timothy/meadow fescue mixture, a Cockle Park mixture, or even permanent grass, according to circumstances or preference. In most cases where conditions normally favour growth, sufficient flexibility can be secured by the management imposed on this main pasture type, and the inclusion of a limited area of Italian ryegrass in the grassland programme. In drier parts it is probably wise to include one special-purpose ley, for instance one based on lucerne with a suitable companion grass. It is the best insurance there is against drought, where irrigation is not feasible.

PLANNING PASTURE USE

Already a considerable amount has been written about the various management processes, apart from top-dressing, which can be employed to arrange a sequence of good grazing—for example, the timing of cutting or grazing, the duration of resting, the planning of pasture establishment, and so on. The important issue is to plan ahead with a real purpose. Herein lies the importance of some subdivision of pasture land, whether it be with temporary or permanent fencing. Only in this way can the potential of grassland be realised.

It has been stressed in the last chapter that a high rate of stocking is a key to a high level of production. But a high rate of stocking makes it doubly important to look ahead so that there will not be a scarcity at critical times. When growth starts to fail there must be reserves to feed ahead of stock—in the form of hay and silage, or grass which is saved *in situ*. If the needs cannot be met by grass in one of its forms, then one must turn to supplementary crops, but it is wise to keep these to the minimum because, in general, they are a relatively dear source of nutrients.

One final point, in securing as close an approximation between the production of grass nutrients and the requirements of stock, relates to timing of calving or lambing. Unless a farm is geared to enter the very limited market for early lamb, it is better to lamb down about a month before active pasture growth commences. The immediate needs of the ewes at lambing time will be met by pastures which have been rested from the end of the previous autumn, supplemented with roots, hay and concentrates. The aim is to have the flock coming to the point of maximum appetite at the time of maximum availability of grazing.

A similar situation holds with summer milk production from pasture, except that it is wise to arrange calving as early as 6–8 weeks in advance of the availability of grazing. The cow calved in late January-February, if she starts her lactation in good condition, can make good use of hay, silage and limited concentrates in the interim before grass is available. Provided a special effort is made to cater for the summer gap, production from grass will be sustained at a reasonably high level till the end of the autumn flush, when the cow will normally be dried off to be wintered mainly on hay and silage.

The April-May calver, on the other hand, may reach a higher peak in milk yield, but invariably her lactation will peter off at about the same time as the cow that has calved two months earlier. April to July are bad months for calving cows which are fed largely on pasture, because invariably they suffer from foreshortened lactations.

You may be thinking, in view of the present relative profitability of winter milk, that it is unrealistic to write about the desirability of deliberately calving cows in the late winter to make the best use of grass. This will not remain the case for very long, partly because the increased production of expensive winter milk well in excess of the liquid market's needs is steadily reducing pool prices and average profitability. We are now at the point where it will pay farmers in the more favourable pastoral districts to aim at a mid-winter dry period for their herd and the fullest exploitation of grazing by using fresh instead of stale cows to utilise the spring flush of growth—more or less in the same manner as the New Zealand dairy industry.

We will come back to this point in Chapter XX where we will be dealing with specific examples of summer milk production. Unquestionably this is a matter of considerable potential importance inside the Common Market because of the very much higher prices that will obtain for cheese and butter and also of cereals required for supplementary feeding. Our thinking is that milk for manufacture will be produced more cheaply under a situation where freshly calved, rather than stale autumn calved cows are utilising grass *in situ* when it is at its nutritional peak.

For the farmer who remains a winter-milk producer, especially in eastern districts, the best calving time seems to be the 6–8 weeks up to the middle of October, because this, among other things, will result in the cow being dry during the mid-summer period. The autumn flush of grass can also be fully exploited, and when the cows are stale at the end of the winter there is the stimulus of spring pasture to level out the lactation curve. Unquestionably, cows calving at this time in the drier counties of Britain have the best lactation performances, and it is from these areas, with their supplies of bedding straw and home-produced concentrates, that we should largely expect most of supplies of winter milk.

CHAPTER IX

FOGGAGE OR COOL STORED GRASS

ALTHOUGH autumn grass lacks the feeding value of spring grass, a good autumn flush is of the greatest importance on stock farms, especially if care is taken in its production and utilisation.

New Zealand farmers, with their highly seasonal farming, have recognised this in their practice of providing what they call "autumn-saved pasture" for feeding in the late winter to freshly-calved cows and to ewes at lambing. In other words, they save grass which would otherwise go to stock at a less productive stage, such as dry sheep or cows near the end of lactation, for periods when they are in full production.

Because of the length and severity of our winter we cannot follow New Zealand practice, except perhaps in more favoured areas such as south-west England and southern Eire. We can, however, go part of the way by making grass available for cattle well into the winter, and by extending the availability of grazing to sheep into the hungry-gap period of February and early

March. But this necessitates a decision to deny good grass to certain classes of stock in late summer, such as dry cows and sheep, so that reserves are created for the late autumn and winter.

The starting point in saving grass is a complete eating out of all rough feed in August and early September. This is necessary for several reasons. The first is that there is a clean start to growth, so that only quality grass will be saved. The second is the creation of conditions which favour desired species, and the third is a turning of this rough feed into dung and urine to stimulate growth.

Generally, some help will be required from nitrogen, applied at rates of up to 100 kg N per ha, according to circumstances. If there has been a warm, fairly dry summer, the need for applied nitrogen will be less than if the summer has been wet and cold. This is especially true if there has been a good growth of clover in the sward during the summer months. Where one is using a pure stand of Italian ryegrass, then a level of at least 70 kg N will be required.

SPECIES FOR FOGGAGE

Italian ryegrass is probably the best grass for back-end grazing, provided it is used before killing frosts set in. It is not a frost-hardy species if it encounters really hard weather when in a lush condition. It will normally stand well until December, and has good milk-stimulating value at this time of the year.

Another pasture which is excellent for back-end grazing is a timothy/meadow fescue mixture, because it maintains both palatability and winter greenness until December. If it is closed fairly early, say by the last week of August, it will normally be ready for grazing by the end of September. If it is eaten out then by the herd, there will generally be sufficient regrowth over the next six weeks to justify another grazing. In fact, a second grazing in the back-end will be necessary with any pasture if there is a tendency for it to get too rank, as this will lead to opening up of the sward.

Though perennial ryegrass throws a lot of autumn growth, it does not stand very well into the winter because of a tendency to go to mush with the first frosts. It is wise to graze perennial ryegrass by about the end of October, especially if it is fairly proud. Any regrowth that is made seems to stand the frost much better than the earlier growth, and it will provide useful December-January grazing for sheep, should these be kept on the farm.

A cocksfoot-dominant sward gives more bulk than almost any other, except Italian ryegrass, when it is laid up for foggage. It must, however, be closed and top-dressed by about the third week of August, especially if S 143 cocksfoot is used, when one is aiming for a full crop. Provided nitrogen fertility is high, it will resist winter burn reasonably well, but after November the proportion of green becomes steadily smaller. By New Year it has a feeding value approximating to that of average hay, and as such can be considered useful food for outwintered cattle, but no more than this. If it is intended to use cocksfoot foggage for milk production, the aim should be to prevent it becoming very proud and to use it before the middle of November, except where climate is very favourable.

VALUE OF LUCERNE LEYS

A few years ago, G. P. Hughes, then of the Grassland Research Station when it was located at Drayton, developed a promising forage-foggage combination in the form of lucerne and cocksfoot, planted in alternative rows, approximately 300–360 mm apart at sowing.

Early in the season, such a mixture will produce either two or three cuts for conservation, according to local conditions. Immediately following the last cut in mid-August, a heavy dressing of nitrogen, say 60–70 kg N, is applied to stimulate a strong growth of cocksfoot which is grazed in the winter by dry stock behind an electric fence. At this time the lucerne will have died back and cattle will follow these rows when grazing, thus minimising damage to the drills of cocksfoot. Lucerne seems to stand poaching very well and, provided the cocksfoot is grazed hard in the winter, the lucerne will maintain a satisfactory proportion in the mixture for a number of years.

This system has not spread as widely as its promise suggested it would, possibly because the growing of lucerne is not generally characteristic of farms with outwintered suckler cows, which are the class of animal which will make the best use of such foggage. Instead, beef men seem to prefer normal pasture mixtures containing cocksfoot for this purpose.

Conventional stands of lucerne, especially those with timothy and meadow fescue as the companion grasses, can also make a very useful contribution to back-end grazing. It is important in promoting winter survival of lucerne to build up root reserves from about the end of August until the cessation of growth, which coincides with the onset of frosts in about the middle of October. At this point, stock may just as well have the lucerne as the frost. There will still be a lot of feeding value in the sward, especially if there is an appreciable proportion of timothy and meadow fescue as companion grasses, both of which retain their quality well into the winter.

PERMANENT PASTURE

A word must be said about the special place which permanent pasture has in providing cool stored grass, especially for a sheep flock at lambing and just before lambing. Preparation is the same as that advocated generally for foggage, except that one need not close the field till October. The aim is not proud growth but fairly dense, leafy material which will stand through the winter without any yellowing at the base of the sward. A number of the usual components of permanent pastures, for example *Poa trivialis* and *Poa pratensis,* are winter-green species and keep their quality very well.

This feed is of special value in steaming up the ewes as they come to lambing. It is so valuable that a farmer is justified in using it on the on-and-off system in February-March, by letting ewes have access to it for about two hours daily, in order to extend its availability.

PLANNING BACK-END GRAZING

Returning to the broader problem of producing and using foggage, it is important to have a fairly clear-cut plan of action so that one can go into the back-end with the knowledge that there is grazing ahead for so many weeks. For instance, a dairy farmer with 60 cows may, with foresight, find himself in the fortunate position, about the third week of September when most of his cows are calving, with 25 ha of good grazing in front of his herd. He will achieve this by concentrating his dry stock on fields which require eating out instead of letting them wander over large areas which are incompletely grazed.

He can then decide a reasonable plan of action to use this grass to best advantage and, with subdivision, he is able to make an appropriate area available to the herd each day. He may, until mid-October, give a fresh break both night and day, but after this date it may be better to keep the cows in at night. The remaining area can then be rationed on a new basis appropriate to the prevailing conditions. If, for instance, the autumn has been a wet one, it may be wise to go over the pastures fairly quickly. But if the ground is dry and the feed is retaining its quality, it might be possible to extend the rationing well into December, with Italian ryegrass as the last pasture to be grazed.

In most years when we were at Cockle Park, an exposed farm, we had good grazing by day for the dairy herd until early December by such planning. We appreciated in full measure the value of this grass, because there was an appreciable drop in milk production when grazing ceased, despite the fact that level of concentrate feeding was increased. This was on relatively heavy land in one of the coldest parts of England.

For those of you who are fortunate enough to farm free-draining soils in a more temperate part of the country, the goal of worthwhile grazing until early December should be easily attained by careful planning. This planning includes more than pasture management, however. It also involves giving autumn calvers a reasonable allowance of concentrates because, as was pointed out earlier in this book, one can over-estimate the value of autumn grass for newly-calved cows if it is the main source of nutriment. Rationed, however, it can, along with concentrates and any other bulk foods that are available, pull its weight by providing valuable protein and, what is equally important, additional variety in the ration.

The effort to produce and use this grass economically is thoroughly justified by the seasonal rise in milk prices at this time of year. So far as sheep farming, or single suckling, is concerned, the essential point is that the need for more expensive supplementary feeding is delayed until the worst of the winter months.

CHAPTER X

HERBAGE PLANTS – SPECIES AND VARIETIES

IN choosing a seeds mixture, a farmer should be as painstaking as is his wife when choosing a hat but, it goes without saying, he should be much more realistic! Like his wife, however, he should not let price be the main criterion.

A seeds mixture composed of cheap commercial varieties proves to be the dearest in the long run, partly because more seeds are required to get the necessary cover and, more important, because commercial types lack leafiness and persistency. Within two years of establishment, pastures which are based on these inferior seeds invariably deteriorate, despite all the management efforts that are made to keep them in heart.

Fortunately, today, the farmer has available at reasonable prices bred varieties of the more important species which are the products of plant-breeding stations in this country and abroad, as well as some genuine old pasture varieties which are produced under reliable certification schemes. There is no need for a farmer who takes any measure of pride in his grassland to purchase inferior seeds. His wisest course is to be guided by the Recommended List of Grasses that is produced annually by the National Institute of Agricultural Botany which undertakes an assessment of new grass varieties as they become available.

The Ryegrasses

Broadly, these fall into two main divisions—short-duration and perennial ryegrasses—but there is an intermediate type first produced by a New Zealand plant breeder, that is known variously as *Grasslands Manawa,* H1, or short-rotation ryegrass. Of the short-duration ryegrasses, the following are the more important:

Westerwolds. This is an annual, which to all intents and purposes behaves like a cereal.

If it is sown in the spring, it will run to head in the summer, and if it is allowed to mature in this way, there will be virtually no regrowth, especially if ordinary commercial varieties are sown, for in these most of the growth is from flower-bearing shoots. Dutch plant breeders have produced more vigorous forms of *Westerwolds* which have greater flexibility, for example, *Sceempter* and *Barenza.*

The great virtue of this grass lies in its quick establishment and the considerable amount of leaf it throws within a few weeks of sowing. On this account, its primary use is as a catch-crop sown at 30–35 kg per ha to give autumn and early spring feed, following an early harvested crop such as vining peas, early potatoes or cereals.

When it is sown fairly late in the summer, about 22 kg of seed can be combined with 90 kg of cereal rye to give the first grazing in the spring. The most that this combination will stand, if quality food is required, is one further grazing, and normally such a catch-crop should be broken in late April or early May, to be followed by a late season crop.

Westerwolds should not be undersown at normal rates in

cereals, for it will dominate the crop. However, a sowing of 2–3 kg along with the cereal is a cheap means of creating useful stubble feed, especially on light land that need not be ploughed till December. This small quantity will not produce enough seedlings to interfere with the cereal crop, but it will produce sufficient seed, which is shed before the corn is harvested, to create a strong growth of seedlings in the stubbles. If this is top-dressed with nitrogen it will give a considerable bulk of grazing in October and November.

Italian ryegrass. This is a biennial, and one of the most useful pasture plants we have, because it is so quick to establish and has such a capacity for out-of-season growth. In addition, it is extremely palatable and it does not run to seed-head as quickly as the hay types of perennial ryegrass. Even when it goes to head, it retains its palatability to quite a remarkable degree. Italian ryegrass is much better as a pasture plant than it is a hay plant, because it loses quality quickly when it comes to full maturity.

Two periods of direct establishment are favoured—either in the early spring to provide summer-gap grazing, back-end grazing and then early bite in the following spring, or in the late summer to provide some light grazing going into the winter but, primarily, to provide extra-early grazing in the next spring. In fact, any grazing of late-sown Italian should be mainly directed towards consolidation and preventing it going into the winter in too proud a condition, for this may impair survival if there are hard frosts. Our preference for such back-end grazing is to put a large number of sheep on the field for not more than 2–3 days at a time.

Normally a properly managed stand of Italian, if a winter-hardy variety is sown, will remain reasonably productive for two years. One can, with reasonable certainty, budget for a second season of early grazing when a stand that has been established in the late summer comes to its second spring. Usually a cut of silage can be taken as well before it is ploughed in.

In districts subject to prolonged frosts, and this includes all the eastern half of Britain, Danish Italian (*Dasas* E.F. 486) or

Combita has an advantage because of its comparative hardiness. Loss of plant during a hard winter can be quite a problem with Italian ryegrass and, as was pointed out earlier, one reason for this is taking a stand into the winter in too proud a condition. Equally, it is dangerous to graze Italian in the back-end in such a way that any recovery growth is removed immediately. Judgement must be exercised to strike a happy medium and if it is necessary to graze Italian at this time then it should be done quickly so that there is no danger of removing this regrowth. An Italian ryegrass that has now become extremely popular is the Belgian bred RvP, possessing a very good combination of yield, earliness, winter hardiness and persistence.

Short-Rotation Ryegrass. This grass, formerly known as H 1, and now as *Grasslands Manawa,* is a hybrid between Italian and perennial ryegrass. It has a finer leaf than Italian, but equals it in palatability. It is capable of vigorous out-of-season growth, especially in the autumn. This makes it particularly susceptible to winter loss, especially where sheep are allowed to graze it continuously going into the winter. Here again, as with Italian ryegrass, the aim should be to allow it to grow leaf to build up reserves, but if leafage is in a dangerously lush condition it should be grazed quickly just before the hard frosts occur. By adopting this type of management with H 1 we have had no difficulty at Cockle Park in maintaining a reasonable plant for as long as five years, but one normally thinks of it as a component of a 2–3 year ley. *Sabrina,* a recent Aberystwyth product, may well fill the same role that *Grasslands Manawa* has had for many years.

Perennial Ryegrass

Though we will concern ourselves here only with bred or certified varieties of perennial ryegrass, unfortunately it has to be admitted that most of the so-called perennial ryegrass sown in this country is a commercial grade which may bear the label Irish or Ayrshire. Very rarely, one may be lucky when buying such seed. For the most part, this commercial seed is of a false perennial type, with a propensity for throwing up seed heads rather than producing the leaf that really matters.

This is a most unfortunate situation, because it tends to bring our most valuable grass species into disrepute. It only exists because the majority of farmers are not sufficiently discriminating to insist on reliable seeds of known origin. We cannot blame merchants for this, except where there is an unscrupulous admixture of commercial seed when certified varieties have been ordered, for their job is to satisfy a demand rather than dictate what that demand should be.

The genuine perennial ryegrasses can be divided broadly into hay and pasture types. This is a somewhat arbitrary classification, for each type is capable of producing either hay or grazing and so the distinction is one of degree. The hay types tend to be rather earlier in growth, more upright in their form, and certainly they are much quicker in throwing up seed heads.

S 24, the hay type bred at Aberystwyth, which is virtually indistinguishable from certified New Zealand perennial ryegrass in its performance, produces seed-heads much earlier than Italian ryegrass, although it does not compare with the latter's earliness in respect of vegetative growth. It is unfortunate that the plant breeders concerned in its development did not secure the Italian combination of early vegetative growth and late ear emergence although *Sabrina* is the nearest variety to fulfill this role.

Today a hay type of pasture plant is really an anachronism, because it belongs to the days of cheap concentrates, when a pasture's function was mainly the provision of bulk. Even when one conserves pasture types of ryegrass as silage or hay, there is ample fibre to suit the needs of any ruminant.

S 23 is a very leafy and persistent ryegrass, it is rather late in starting growth in the spring and it is not markedly palatable. It is particularly useful for upland reseeds, where S 23 is able to persist under conditions of relatively low fertility.

An alternative is genuine old pasture Kentish indigenous ryegrass, which is also earlier than S 23 in vegetative growth, extremely persistant and, in our experience, a more palatable grass which seems to thrive on really intensive stocking.

In the last few years new Continental varieties have to a very large extent become available to compete with the S varieties from Aberystwyth. At one time, S 22 Italian ryegrass was the most used of Italian but now RvP is probably the most popular with leading farmers, although reports of rust susceptibility from some areas may be a warning against using too much of this variety.

The National Institute of Agricultural Botany lists Recommended Varieties of Grasses has two recommended Aberystwyth perennial ryegrasses, S 24 early type and S 23 late type. S 101 referred to in previous editions of this book is now outclassed, while S 321, a medium late or intermediate variety is not recommended.

For longer leys and particularly for intensive use, such as dairy and sheep paddocks, the more prostrate pasture types are recommended. S 23 is still used to a very large extent but a Belgian variety, *Melle,* has shown tremendous ability for persistence and high production under these conditions. In our experience it is a more palatable variety than S 23.

The development of varieties of tetraploid perennial ryegrass, with their high yields and broad leaves, has resulted in such varieties as *Reveille, Barvestra, Taptoe* and *Petra* being increasingly used. Three factors must be borne in mind with these. First, the seed is large and comparatively higher seeding rate is required at establishment, although this may be partially balanced by the seedlings from these large seeds having a higher survival rate. Second, yields of these varieties may look deceivingly high and the lush wide leaves may seem to out-yield normal diploid varieties. Third, they have an appreciably lower dry-matter content, a factor which is important and with direct cutting of grass for silage, as a low dry-matter product may result.

The fairly common practice of sowing late pasture and early hay types of ryegrass together in a sward, for example, S 23 and S 24, is not recommended. One seems to get a succession of seed heads thrown up over the whole of the summer period, and ryegrass is difficult enough to manage in this respect without aggravating the problem. If any blending of varieties is undertaken it should be within types, for instance S 24 and *Grasslands Ruanui* or S 23 and *Melle.*

Grasslands Manawa and *Sabrina* ryegrass, which also tend to have a late ear emergence despite the earliness of their vegetative growth, may advantageously be included in ryegrass mixtures, particularly with early or mid-season varieties like Kent and S 24. It will give a longer period of vegetative growth, especially in the first two or three years of a pasture's life.

Timothy

Formerly, timothy was regarded mainly as a hay type of grass with an adaptation to strong, moist land. Today, it is accepted by progressive farmers as one of our most valuable grazing species, suited to a wide range of soils provided fertility is high. This has come about through the increasing popularity of the timothy/meadow fescue mixture up to the late sixties which, in turn, has been made much more valuable because of the qualities of S 48 timothy, possibly the best grass developed at Aberystwyth.

Timothy is exceptionally leafy and productive, particularly during the early summer when most other species are tending to run to head, and it is among the most palatable of our long-term pasture species. It is not a vigorous autumn producer, but the herbage stands well into the winter without deteriorating. Generally speaking, it is a grass which gives of its best under a system of on-and-off rather than continuous grazing.

These attributes of timothy apply particularly to S 48, and to some of the new bred Continental pasture varieties. Very little can be said in favour of commercial North American, Scottish or

Swedish timothies, which throw a lot of stem and do not produce a good aftermath. Their only conceivable use is in a 2-year ley for hay, but few farmers will want to sow such highly specialised pastures since hay should normally be a by-product rather than a main product of pasture farming.

The earlier Aberystwyth variety, S 352, is a much more persistent and valuable cutting grass than these commercial varieties, and it is an ideal companion for lucerne. It lacks the grazing qualities of S 48 although it is not without virtues in this respect. One would not recommend it being used as a substitute for S 48 in a mixture which is intended mainly for grazing.

A word must be said about a third Aberystwyth timothy, S 50 which is prostrate and spreading in its habit of growth, and very winter green. Unfortunately, it produces very little seed and therefore it is expensive to buy. Otherwise it could be a very useful grass in long-term grazing pastures, especially on the uplands. The most one can suggest for it, is that 1–2 kg of seed might be included in a mixture for such conditions, but even so it is doubtful whether there will be a worthwhile establishment in competition with a strong-tillering type of ryegrass like S 23, which should be a main component of such a mixture.

Meadow Fescue

This is an indigenous species which is a not very conspicuous element of old pastures, especially on heavy land. It is a fairly deep-rooting perennial with a wide adaptation on fertile soils. The leaf blades are not unlike Italian ryegrass in appearance, and it makes appreciable growth early in the season. It is second to timothy in respect of palatability among the long-term pasture plants, but it is earlier in both vegetative and reproductive growth than timothy, for which it is an ideal companion in association with clover in grazing leys. It has a good aftermath performance, in that it throws leaf after cutting rather than seed-heads, and it also had a good autumn growth, which is made more valuable by the fact that it remains relatively winter green.

The two Aberystwyth-bred varieties, S 53 and S 215, are still on the N.I.A.B. Recommended Lists, but are often replaced by Continental varieties. S 215 is ousted by a Dutch variety, *Comtessa,* with rather higher yields in later summer and autumn. The late varieties recommended are again Continental in origin, *Bundy* and *Admira* pasture meadow fescue from Holland.

Meadow fescue is slow to establish and one does not see its real merit in the establishment year. It is therefore recommended when sowing timothy/meadow fescue ley that some 5–6 kg of an Italian ryegrass be included in the mixture to give higher production without detriment by competition in the first year.

Cocksfoot

Cocksfoot is one of our deepest rooting grasses and, judged purely on a dry-matter basis, it is also one of our most productive species, especially on soils which are inclined to suffer from summer drought. Its value would be even greater if plant breeders could produce varieties with a greater palatability and higher digestibility. This does not deny that they have not made very considerable progress with this species. The named varieties are a great improvement on the old commercial varieties, which are coarse-growing and stemmy and lack persistency if subjected to hard grazing. They are also inclined to burn very badly, going into the winter.

Commercial Danish belongs to this category and its use is not recommended, despite the fact that it is about the earliest cocksfoot to show growth in the spring. If the need is for an early-growing cocksfoot it is advisable to use a named Scandinavian variety such as *Roskilde.*

Aberystwyth has produced many varieties of cocksfoot—S 37, which essentially a hay type and a second-early rather than an early variety; S 26, a rather leafier dual-purpose cocksfoot; and S 143, the "mop" cocksfoot pasture type which stands fairly

PLATE 9a

Eighteen-month-old Friesian steers winter fattening in yards on a diet of mainly hay and self-feed silage, following a summer on grass.

PLATE 9b

It is important to use breeding cows which are adapted to the conditions of the farm. The hardy breeds such as the Galloway and the Welsh Black are more suited to poor upland grazing than earlier maturing breeds.

PLATE 10a

A home-made heavy roller which will follow the contours of the ground. This is extremely useful for summer reseeds, as it shatters surface clods and gives a really firm finish—a vital factor in securing successful late-season establishment.

PLATE 10b

Cockle Park Jerseys grazing behind an electric fence.

hard grazing and is a valuable component of a Cockle Park mixture or of a mixture with timothy and meadow fescue.

Other Grasses

There has, over a number of years, been some interest in tall fescue. Aberystwyth S 170 is the best variety available which is rather coarse and unpalatable, except in the winter and early spring when stock will graze it if there is nothing else they like better. In New Zealand tall fescue is regarded as one of the worst of the weed grasses, especially on heavy land, because it is so aggressive. It may be that some American or Mediterranean varieties, or S 170, may have some value ultimately, but at this stage it is probably wise for a farmer unless he is a particularly good grassland manager to forget about tall fescue and concentrate on grasses of established merit.

On poorer soils crested dogstail is probably a more useful bottom grass than is generally realised, and it is worth including in seeds mixtures for marginal land likely to be subject to very hard grazing by sheep. Creeping red fescue (S 59) may also be of some value under similar conditions, and while it is not a high producer, it is winter-green.

Legumes

White clover. This is given pride of place under British conditions for reasons which have been sufficiently outlined in Chapter V. Two main types are recognised, the long-petioled white clovers, of which the best known example is the Aberystwyth-produced S 100, and the more prostrate creeping type of white clover exemplified by Kent wild white clover, an ecotype which is the product of centuries of intensive grazing on Romney Marsh. Aberystwyth has produced a prostrate variety, S 184 which is similar to Kent.

New Zealand white clover or *Grasslands Huia,* either mother seed or pedigree, closely resembles S 100, and from a farmer's viewpoint is a completely acceptable alternative, the only governing factor affecting choice being that of price. Both are extremely persistent under mowing and grazing conditions, provided the grazing is not hard and continuous. If there is a likelihood of this happening, as for instance on an intensive sheep farm, then it is wise to include in the mixture a small quantity of Kent wild white clover, say 0.5 kg per ha, along with 1.5 kg of S 100 or *Huia.*

This will not generally be necessary with dairy pastures. The tall-growing clovers, which are much quicker to establish than wild white, are able to hold their own very well with competing grasses, provided the balance of power is not upset by repeated dressings with nitrogenous fertilisers.

Ladino white clover is even stronger growing than S 100, but it lacks persistency if it is grazed. It is not recommended for British conditions because it is not sufficiently versatile in its behaviour. A much better prospect is Kersey white, which is also a strong growing clover characterised by good growth early in the season and a good recovery after cutting.

There is no real place for Dutch white clover in our agriculture, outside of one-year leys, because it is no more than an annual. The danger is that the ignorant or the unscrupulous may include it in a long-term mixture. A particularly interesting new introduction is the Welsh Plant Breeding Station's varieties, *Sabeda* and *Olwen*, which have given very good results, due to their long petioles, in the west side of the country. The current marked increase in the price of fertiliser nitrogen may necessitate a greater reliance on clover nitrogen and this variety could be of considerable value in this connection.

Red clovers. These are also of two types—broad red clover (sometimes called double-cut cow-grass), and late-flowering red clover which will persist in a ley for 2–3 years.

Broad red is the usual companion of Italian ryegrass in one-year leys for mowing. Apart from the bred variety, S 151, there are several good local varieties, for example, Essex, Dorset Marl and Vale of Clwyd. Not only do these clovers make a big contribution at their first cutting but they also produce a strong aftermath which may be cut for hay, taken for seed or used for grazing (but with precautions because it is liable to cause bloat).

Red clovers are subject to so-called clover sickness, which is due to a fungus (*Sclerotinia*). Where this is serious it is necessary to substitute alsike for broad red or late-flowering red clover in short-duration leys.

A new development in red clover varieties is the tetraploid broad red varieties, *Hungaropoly, Red Head, Teroba* and *Tetri,* the first bred in Hungary and the other three in Holland. Their outstanding advantage is their high resistance to clover rot and their very high yields, and their relatively high persistency resulting in substantial yields in their second year. Again with higher nitrogen fertiliser prices, these varieties will no doubt be used increasingly with Italian ryegrass to produce very high yields of herbage, particularly for cutting for silage, without the use of massive applications of nitrogen. The Aberystwyth variety, *Sabtoron,* has high levels of resistance to stem eelworm and it persists with good management up to three years from seeding.

Sometimes, and more particularly in the north of Britain, broad red clover is included in a long-term ley of a Cockle Park type in order to provide bulk in a hay crop that is taken in the first harvest year. It is not a practice that we can recommend in our strong belief that the management of any long ley in its first year should be directed towards building up of fertility and promoting the establishment of perennial species.

Among the most valuable of the late-flowering reds, which are more useful for grazing than broad red clover, is the Montgomery type red, which may be home-produced or of New Zealand origin. Aberystwyth has produced a bred variety, S 123, which is excellent for grazing. The great value of the Montgomery red is in 2–4 year leys which are used for cutting and for grazing in an on-and-off system. It does not persist with hard continuous grazing, but is a useful contributor to aftermath, especially in drier districts, where its deep-rooting habit is advantageous. Late flowering red clover is not, however, a particularly palatable plant, even though weaned lambs thrive well on it in a dry summer.

Other Legumes. Alsike clover, the seed of which mainly comes from Canada, does not compare with a good white clover where the latter thrives, but it has some value on damp, acid land under conditions of fairly low fertility. As was pointed out above, it is resistant to clover sickness and may be used to replace red clover where this disease is serious.

Sainfoin still retains some popularity on chalk or brash soils. There are two types—giant sainfoin establishes rapidly but lasts for only 2–3 years, while common sainfoin usually reaches peak production in its third year, provided there is no serious invasion by wild grasses. It can be used for cutting or grazing, but it is important that it should have adequate rests after each grazing if it is to survive. In this respect it is very like lucerne. It is extremely palatable, but because of its strong stems it is not easy to make good sainfoin hay in broken weather. Establishment may be made directly, using 45 kg of milled seed per hectare, or it may be undersown in a cereal crop.

English trefoil, or black medick, is an annual which is sometimes undersown in a cereal, especially on chalk soils, to provide stubble feed for sheep before its residues are ploughed in at the end of the autumn. Its usefulness is limited to this function, which includes some contribution to soil fertility as well as sheep feed. It should not be used as a substitute for red clover in a one-year ley.

While subterranean clover is indigenous, especially near the sea coast in the south of Britain, it has never achieved any agricultural significance in this country. It is, however, an extremely important legume in areas of low summer rainfall in the southern part of Australia and on dry soils in New Zealand. It is an annual which buries its seeds, which germinate with the autumn rains to produce a strong growth which persists through to the early summer. The dry residues which remain have quite a high feeding value. This plant has made a wonderful contribution to livestock production under difficult conditions, but it has no advantage over white clover where this species can be grown.

Crimson clover, or trifolium, is now more or less a novelty. It is an annual which is generally sown on a cultivated stubble to provide green fodder for the mid-spring period, either alone at 17–22 kg per ha or in association with 10 kg of Italian at a rate of about 14 kg per ha. It is not a particularly palatable plant, and it has little to recommend it, for one can get more value out of a straight seeding of Italian following an early harvested cereal crop.

For those who feel *Herbs* are a desirable component of leys, the following list is given without comment, along with suggested seed rates if they are incorporated with a mixture of grass and clover seeds. Double rates should be used along with 3 kg of certified white clover if they are sown in a separate herb strip.

Chicory $1\frac{1}{2}$–2 kg/ha
Burnet 1–$1\frac{1}{2}$ kg/ha
Ribgrass or plantain 1 kg/ha
Yarrow 0.15 kg/ha
Sheep's parsley 1 kg/ha

This recipe is given with reservations because there is no substantial evidence that herbs should be deliberately included in a grazing mixture.

CHAPTER XI

GRASS SEED MIXTURES

THERE are few grounds for dogmatism in any discussion on grass seed mixtures. In part, this is due to a paucity of reliable evidence on the comparative advantages of the various species combinations. But the main reasons lie in the diversity of conditions of the demands which are made upon them.

In making a choice of seeds, as in taking so many farming decisions, it is very much a question of horses for courses. Victorian farmers believed that there was safety in numbers, not only in respect of the size of their families but also in respect of ingredients in seeds mixtures, which usually included a little of practically everything in the seedsman's catalogue. Since that time, however, there has been a marked swing towards simplification.

This rationalisation was pioneered by Gilchrist, whose Cockle Park mixture, still one of the most popular, was originally based on three grasses—timothy, cocksfoot and ryegrass—and three legumes—red and white clover and trefoil. The modern school has carried simplification further, and even advocates ultra-simple mixtures based on only one grass and sometimes no clover.

However, within these very simple mixtures diversity is secured by the inclusion of varieties of a species, for example, S 23 and *Melle* ryegrass together with S 100 and S 184 white clover. Any further complexity, either for good or for ill, comes from such volunteers as *Poa trivialis* and common bent, which are invariable invaders of sown grassland.

PRINCIPLES IN COMPOUNDING MIXTURES

The main considerations in compounding a seeds mixture are as follows:

(a) cost of ingredients
(b) compatability of species
(c) intended life of the pasture
(d) environmental conditions
(e) function of the pasture.

Cost of ingredients becomes less important as the intended life of the pasture becomes longer. If, for instance, one is sowing down Italian ryegrass merely to provide stubble grazing, then a cheap commercial lot, provided germination is satisfactory, will be a better proposition than a more expensive bred variety like S 22 or RvP, which is fully justified when the pasture is intended to last into a second year. When it comes to longer leys, cheapness of ingredients is surely a very false economy, if obtained at the expense of leafiness or persistency.

The difference in price between good certified seeds and doubtful commercial categories seldom amounts to a large amount of money. It is very easy to be a penny wise and a pound foolish, especially when one remembers that some commercial strains demand an appreciably higher rate of seeding. The certified strains have been selected for their tillering capacity, and so it is possible to economise by seeding at comparatively low rates without detriment to sward density.

Compatibility of species depends mainly on such factors as relative tolerance to light and shade, rapidity of establishment and reaction to intensity of defoliation. Meadow fescue is a very unsuitable ingredient for a pasture which contains perennial ryegrass, because it is slow to establish. Invariably it is swamped completely by even relatively light seedings of the bred varieties of perennial ryegrass, which are very prolific in the production of tillers. Quite the best companion for meadow fescue is timothy, while possibly cocksfoot can be used. All three of these grasses are slow to establish, so there is little danger of one species smothering the others at the seedling stage.

Considerable skill in management is necessary if these grasses are to be kept in balance after establishment. Cocksfoot can easily become dominant with lax grazing, or repeated closing of the field for hay or for foggage. Because it is generally less palatable than either timothy or meadow fescue, cocksfoot is less severely punished where there is continuous grazing which does not effect complete utilisation. On the whole, all three grasses tend to do better under a system of on-and-off grazing, such as obtains with close sub-division and rotational grazing or folding using the electric fence.

Usually the aim should be for a longer interval between grazings with swards based on these three grasses than with a dominant ryegrass pasture, which gives of its best with shorter intervals between grazings. One of the big problems with ryegrass is that of keeping on top of it to minimise the production of seed-heads, and in the process to ensure that there is sufficient light to encourage white clover.

VALUE OF COCKSFOOT

Cocksfoot is very often sown in association with perennial ryegrass, as for instance in the modern Cockle Park mixture:

8 kg cocksfoot
14 kg perennial ryegrass

5 kg timothy
3 kg late flowering red clover
2 kg white clover

The cocksfoot is not very noticeable in the first year following establishment, and indeed it will never achieve dominance if it is always subjected to hard continuous grazing, but converse treatment will swing the pendulum in its favour.

Of the several varieties of cocksfoot available, S 143 is probably the best to sow with ryegrass. This is the "mop" type which is prostrate in growth habit. and one would expect it to stand up to the sort of management one accords to ryegrass rather better than the taller-growing S 26 or S 37. This has been our experience in managing cocksfoot/ryegrass mixtures. But we must admit that this experience is no longer recent, because of a now very marked preference for planting the species in association with timothy and meadow fescue, whenever circumstances justify the sowing of cocksfoot. Unless there are vey special circumstances, such as thinness of soil and liability to drought, this combination is best sown in fields which are required less for milk production or fattening than for growing store stock. This argument is based on the relative lack of palatability that characterises cocksfoot.

Invariably, in sowing this mixture, it is recommended that 5–6 kg of Italian ryegrass should also be included to provide higher production in the first harvest year. The following is suggested as a suitable mixture of this sort:

6 kg RvP or *Dasas* Italian ryegrass
7 kg of S 143 cocksfoot
7 kg of S 215 meadow fescue
6 kg of S 48 timothy
2 kg of S 100 or *Grasslands Huia*

Neither of the two ryegrasses suggested has aggressive tillering habits, and at the rate of seeding proposed there is little danger of impairing the establishment of the remaining species, always providing that the ryegrass is tightly grazed in the first few months following sowing.

These remarks apply especially where establishment is by direct reseeding. Here a danger to be guarded against in the seedling stage is an unwanted dominance by annual weeds. A quick-covering grass like RvP or *Dasas* Italian ryegrass helps to control weeds because it is possible to stock the field so much sooner after sowing. This point is referred to in greater detail in the following chapter dealing with pasture establishment.

Turning now to environmental conditions, some reference has already been made to this when discussing one attribute of cocksfoot, namely its considerable capacity to stand up to drought. For those farmers who have a high proportion of gravel soils, cocksfoot must be the basis of ley mixtures, either in association with other grasses and clover or with lucerne, which, with its very deep rooting system, is the best drought resister we have.

TIMOTHY AND MEADOW FESCUE

Many farmers prefer timothy and meadow fescue to cocksfoot as a companion for lucerne, and our experience leads us to agree with them. In the first place their palatability helps to increase the attractiveness of the mixed herbage and, secondly, they are not so aggressive as cocksfoot. Even at a seeding as light as 2 kg per ha, cocksfoot can, in a matter of two or three years, become dominant in the sward. When this happens the ley becomes less valuable in a really severe drought, because cocksfoot cannot compare with lucerne under such conditions. Timothy and meadow fescue are less competitive with the lucerne, and though they themselves will contribute little in a drought this is more than compensated by the bigger contribution from lucerne.

The traditional view of timothy as a moist land species has probably operated unjustifiably against its more widespread use. It is, in fact, one of the most useful species on comparatively light soils such as chalk. The same applies to its usual companion, meadow fescue. These two species in combination with white clover, constituted for many years our most valuable milk producing ley until the more recent move to leys based on pasture types of perennial ryegrass.

A good mixture for milk production, and we like it, too, for fat lambs, is one that combines all three grasses together with white clover:

6 kg RvP (or *Manawa*)
10 kg *Sceempter* pasture meadow fescue
7 kg S 48 timothy
2 kg S 100 white clover or *Grasslands Huia*

The three grasses of this mixture dovetail in their growth pattern. The earliest part of the flush is provided by RvP followed by the meadow fescue, which can be regarded as a second early, while timothy makes its greatest impact in the late spring, being leafy when most of the other species are running to head.

One of the great virtues of this mixture, thanks to the timothy, is that it can be closed as late as the first week of May, after at least two grazings, to secure a heavy and fairly leafy crop of hay. Following cutting, meadow fescue and clover will give a good summer aftermath, which will be followed in turn by a strong autumn growth of RvP.

We have had many leys of this type at Cockle Park and they have been maintained with a dominance of the sown species for up to six years. They have been the highest-producing pastures on the farm though admittedly they have been nursed to avoid the effects of poaching or continuous over-grazing. Fertilisers have not been spared for these are high-fertility-demanding species which give of their best when quick grazing or cutting is alternated with reasonably long rest periods. Pastures based on this mixture are not good for out-wintering fields because they tend to be fairly open and are liable to weed invasion.

VARIETY OF SWARDS

If the policy is one of hard continuous grazing, ryegrass is the most satisfactory species for such treatment and it is due to its ability to respond to intensive systems and high stocking rates that ryegrass mixtures are now the most widely used. Perennial ryegrass is one of the best grasses in respect of recovery after poaching. The timothy/RvP/meadow fescue mixture (which we call a four flusher, but not in the American sense) gives a ley which can only be a sacrifice field once in its lifetime, that is immediately prior to its being ploughed out. But a ryegrass ley, after it has been established to the point that there is a dense turf, will show remarkable powers of recuperation and probably this is another reason for its increase in popularity over the past few years.

Generally speaking, and this has been mentioned once before, it is unwise to get too many sorts of ley on a farm because management has to be so specific for each type, and there is the complication that silage clamps become a mixture of too many sorts of herbage. Diversity of leys may become necessary if there is a diversity of soil types, but where soils are reasonably uniform, leys could be based on the following mixture:

6 kg RvP Italian ryegrass
12 kg S 23 or *Melle* pasture perennial ryegrass
6 kg S 48 timothy
2 kg *Grasslands Huia* or S 100 clover
0.5 kg Kent wild white clover

Though timothy will not make much contribution in the early life of the sward, it will have an appreciable part to play later, especially if the ley is used for conservation.

Such a programme of leys does not preclude having some Italian ryegrass in addition, especially on dairy farms where really early bite is required. This may be no more than catch-crop Italian undersown in a cereal crop without clover at the rate of 25 kg per ha. Such a pasture, after one or two grazings and perhaps a silage cut, might be broken for rape or late-sown kale.

ONE-YEAR LEYS

For one-year seeds undersown in a cereal crop and intended for cutting the following mixture is suggested:

18–24 kg Danish *Dasas* EF486
5–7 kg broad red clover or tetraploid red clover e.g. *Hungaropoly*

Usually this mixture will provide no more than stubble grazing, a main cut of hay, and a clover aftermath which may be grazed, cut either for hay or seed, or ploughed in to provide green manure. Where clover sickness exists, the tetraploid clover must be used. Where stem eelworm is a problem the resistant variety, *Sabtoron*, may be used.

With this exception, no special hay mixtures are advocated. Except for very extreme pasture types such as S 50 timothy, which is not really a commercial proposition in any case, one gets enough fibre and bulk from pasture types without sowing hay varieties—unless, of course, one is growing grass merely to sell hay. Hay types belong to a past when concentrates were cheap, not to the present, where we are more concerned with leaf than with stem.

SUITING THE SWARD TO FARMING NEEDS

In one of the seeds mixtures given above, 0.5 kg of Kent wild white clover was included. It could have been S 184, also a prostrate type of clover which stands up well to hard grazing. The less prostrate clovers like *Grasslands Huia* and S 100 can be lost from a sward with severe grazing, such as one may get on a sheep farm, but genuine wild white clover is remarkably resistant to this treatment. In any mixture intended mainly for sheep grazing this small admixture of wild white is an insurance. You may think that this is a ridiculously low seeding rate, but remember that 0.5 kg of white clover contains at least 700,000 seeds, which spread evenly over a ha, gives 70 seeds per square metre.

This is an example of fitting a seeds mixture to an environment. Another is the occasional inclusion of alsike clover on stiff clays which are slightly acid and where experience shows that it is not easy to establish white clover. Generally, however, if it is possible to plough land to establish a pasture, then it will be equally feasible to apply lime and phosphate to provide the conditions where white clover will flourish.

This raises the question of seeds mixtures for difficult conditions, such as upland reseeds. Here it is stressed that there is seldom any need to sow such species as Yorkshire fog, *Agrostis* and sheep's fescue. Usually there is sufficient seed in the soil for these grasses to establish should it not be economically feasible to maintain the level of fertility which better species demand.

On this land, which cannot be regarded as being suited for intensive grassland farming, there is a lot to be said for a modified Cockle Park mixture of the following nature:

6 kg Danish Italian *Dasas* EF486
14 kg S 23 or Kent ryegrass
6 kg S 143 cocksfoot
4 kg S 48 timothy
2 kg crested dogstail
2 kg S 100 white clover or *Grasslands Huia*
1 kg Alsike clover
0.5 kg wild white clover or S 184

The seeding rate is higher than that usually advocated for a lowland pasture, but this is a recognition of the fact that it is seldom possible to create a really first-class seedbed. Ideally, of course, such a seeding should be made after a pioneer crop such as rape and Italian ryegrass so that any mat or peaty layer is thoroughly broken down and incorporated through the soil. The 6 kg of Danish Italian has been added to the mixture to act as a grazing nurse, enabling stock to be put on as quickly as possible after seeding to effect essential consolidation.

CERTIFIED SEEDS

It will be noted that in the above mixture, except for the dogstails and the Danish Italian, relatively expensive certified seeds are prescribed. This is deliberate, because the emphasis must be on persistency—one cannot afford to plough upland fields of this type very often. If persistent species are sown, there is a reasonable hope of keeping them dominant by maintaining fertility at a satisfactory level. Usually on such land this will mean lime and phosphate application.

There is one last point to be made. It is useless sowing any valuable seeds mixture unless there is commensurate management. If you are not prepared to provide the fertility and the management which the high-producing species demand, you may as well accept the cheapest commercial mixture that your merchant can provide. It will almost certainly contain a lot of inferior material, but you must not complain, because your merchant will be providing something that matches your grassland farming.

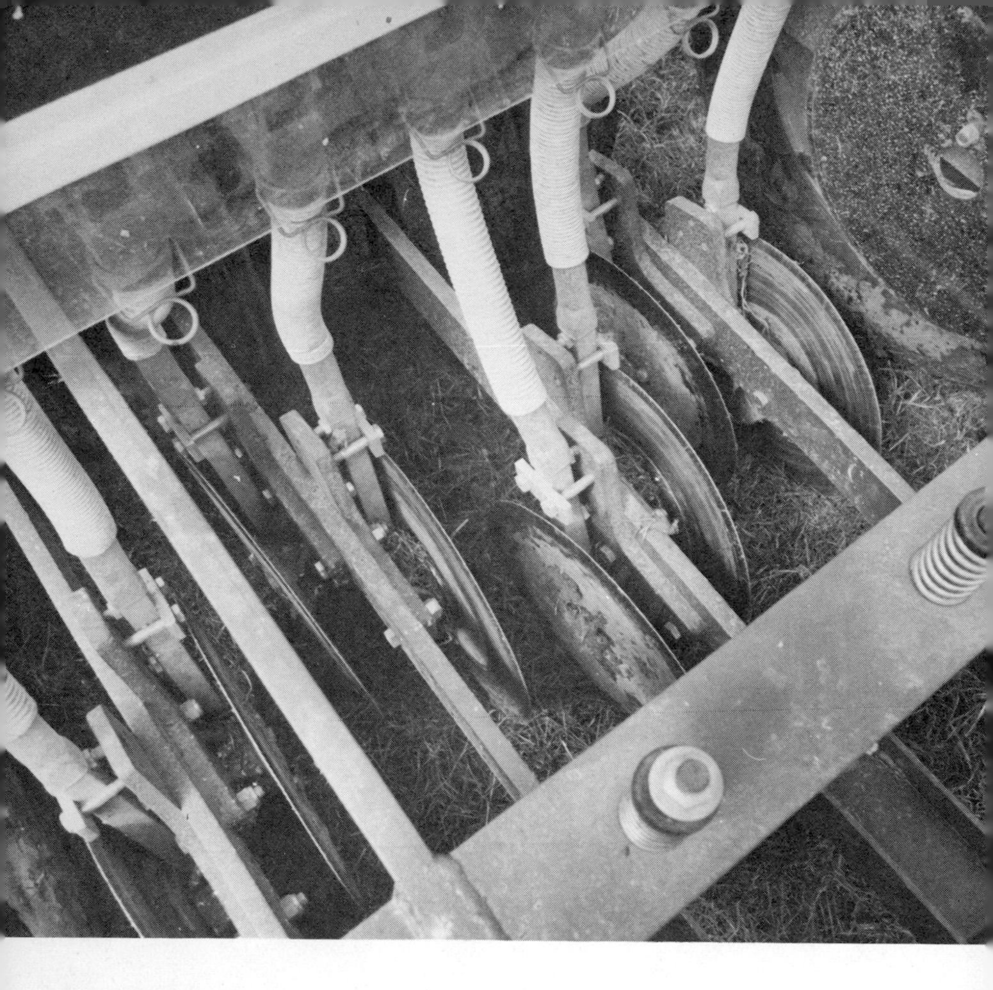

PLATE 11a

Close-up of the seed coulters and cutting discs on the front-and-rear-sprung drag arms of the Moore Uni-drill. The discs open up a slot for the seed to fall into, which allows direct reseeding without disturbing the rest of the pasture. Although primarily a seed drill, the Uni-drill can also be used to direct drill cereals and crops such as stubble turnips and kale.

PLATE 11b

With chemical ploughing, using one of the new materials to destroy grass, a reasonably clear surface can be prepared. Over-sowing with grass and clover seeds can then be carried out with a sod-seeding machine such as the Howard Rotaseeder, seen here.

PLATE 12

A creep gate with adjustable second-from-bottom bar allows lambs access to good grass while controlling the ewes.

CHAPTER XII

PASTURE ESTABLISHMENT

THE two principal methods of establishing a pasture are (a) undersowing of a cereal crop intended for harvest, and (b) direct reseeding with or without a grazing nurse crop which may be a cereal or a cruciferous crop like rape. A third, less usual, method which is in the nature of a compromise between these two is the undersowing of an arable silage mixture.

Of these methods quite the most popular, and equally the most suspect, is the first—namely the undersowing of a cereal crop. It is popular because it is cheap, in that the major costs of cultivation are carried by the cereal crop, and there is also the feeling of having one's cake and eating it, in that pasture establishment is a by-product of cash cropping.

The latter part of this argument is not without flaws. In the first place, there must be some sort of compromise between the needs of the corn crop and the needs of the establishing pasture. Usually, one adopts a lower rate of seeding and fertilising for the cereal with the result that the crop is below the potential of the land.

This is an important consideration, because yield is lost at the profit end of the operation. However, this is usually a minor consideration as compared with a much more serious loss arising from poor establishment of pasture, either because of competition for moisture in a dry season, or because of lodging and mechanical and other damage which may occur in unfavourable harvesting conditions. This danger is not so acute if one is sowing a cheap seeds mixture which is intended to have a very short life, but it may have serious consequences for a long-duration ley based on expensive seeds. Here there is usually a temptation to persevere with a half-failure in the hope that it will fill up. Usually it will fill up, not with sown species, but with volunteers which will impair the subsequent productivity of the pasture.

ARGUMENTS FOR DIRECT RESEEDING

The strongest argument advanced by the protagonists of direct reseeding is that there is a much greater certainty of getting a full establishment of the sown species. Certainly this is true if the job is done properly by attending to essential points in pasture establishment. They answer the objection that there is a loss of the grain crop by pointing out that there is a counter credit of valuable grazing in the establishment year of a spring reseed.

Provided it is sown before mid-April, there will be fresh grazing during the summer-gap period from June to August, when older pastures are losing quality. It is well recognised that there is no better stimulus to milk yields at this time of the year than a spring-established pasture. By the end of September such a pasture can quite easily have produced 4,500 litres of milk to the ha, a substantial credit to set against the loss of a grain crop.

This raises another consideration affecting pasture establishment policy, namely the type of farm in question. If the farm is large and grass use is extensive rather than intensive, as it is under normal conditions of fat lamb and beef production, the case for direct reseeding is not as strong as it is in the small dairy holding, where every effort is being made to make the most of grass by heavy fertilising and intensive grazing methods.

Not only is a spring reseed more valuable to the dairy farmer than it is to a fattener, but there is the subsequent consideration that intensive methods of managing pasture are only fully justified if the pasture is a good one. In other words, the more efficient a farmer becomes in using his grass, the stronger becomes the case for establishment by direct reseeding.

LATE SUMMER RESEEDING

One very cogent argument against direct reseeding in the spring is that land may be out of production for such a long time. If the previous crop is a cereal, this period can extend from August through to May. Even if undersowing is somewhat faulty, at least it will give some useful stubble feed in the autumn, especially if it receives help in the form of applied nitrogen. It will also give keep in April, which on most farms seems to be the hungriest month of the year.

This counter-argument only applies if one follows the practice of direct reseeding in the spring. There are, however, the strongest grounds for advocating direct reseeding in the late summer as an alternative to under-sowing or spring seeding. In the southern part of Britain, except on farms which are late because of their elevation, there is usually ample time to follow an early-harvested cereal crop with a direct reseed and to get a first-class establishment before the onset of winter.

Such a policy of pasture establishment has a great deal in its favour. In the first place one can aim at a full cereal crop without any fears about prejudicing the undersown seeds. Secondly, because the soil is warm in August, the grass and clover seedlings develop very quickly—in most autumns there will be a really worthwhile flush of growth within a month of seeding, and there will be vigorous growth well into the back-end. Thirdly, these late-summer established leys are especially vigorous in the following spring, and so make a full contribution at a time of normal food shortage. On all scores, except that one loses the summer milk stimulus of the spring-sown ley in its establishment year, the case for direct reseeding in the summer, especially with a ryegrass ley, is so strong that it is surprising that the practice is not more widespread.

Two main arguments are advanced against late summer seeding. The first is that it is another job to clash with harvest, but rarely do we have a harvest season that is not interrupted by broken weather which gives the opportunity for the necessary cultivation.

The second objection, refers to the difficulty of getting a satisfactory take of clovers and this can be very real if precautions are not taken. It is essential to let light into the sward, both in the autumn and again in the following spring. This latter period is especially critical, and on no account should one take hay or early silage from the field, though a cut of late silage is permissible if it is in the interests of the farm.

In our experience late-summer reseeding has always been satisfactory when sowing has been completed by the end of August, but there is a real risk of partial failure and with it a necessity for spring patching with late September seeding, especially with heavy soils.

In Kent, the reseed usually followed barley or winter oats, which are harvested earlier than wheat. Occasionally it followed canning peas, which in many respects is an ideal preceding crop because the land is easily worked and is in very good heart, while in most seasons cultivations can be completed before cereal harvest begins. The only shortcoming of this succession is the danger that a residual infection of pea weevil may impair clover establishment.

Unfortunately, in Northumberland the cereal harvest is so late that there is insufficient time for reseeding after cereals. Here we evolved a system of July-August reseeding after Italian ryegrass which has been established in a cereal crop in the previous year.

This, in a sense, represents a compromise between two recognised methods of establishing pasture. Because only 24 kg of Italian ryegrass are sown per ha, there is relatively little at stake if the fully fertilised cereal crop lodges, or if there is mechanical damage from heavy equipment during a wet harvest. No clover is sown with the ryegrass so it is possible to adopt normal spraying for weeds if required.

Of course, this practice necessitates the use of nitrogenous fertilisers on the ryegrass to get full production from it. In most years an application of 40–55 kg per ha of nitrogenous fertiliser immediately after harvest gives a good flush of stubble feed, which is especially valuable for wintering in-lamb ewes. A further dressing in the early spring will give feed that takes the burden off long-term leys which are allowed to retain leaf at a critical time of the year. Usually towards the end of April further nitrogen is applied, and the ryegrass is then closed for a silage cut which is taken at the beginning of June. Nearly two months intervene before reseeding, and this period can be used for a half-fallow to destroy couch grass and other weeds.

This approach to reseeding has been a very successful one from a farm management view-point, because we have an area for out-of-season grazing which we don't mind poaching because it is in a cheaply-established ley. Furthermore, both our summer-established long leys and the Italian are there to help us through the critical month of April, when all the mouths on the farm are hungry mouths. When the land is out of production in June, July and August, there is less grazing pressure on available pastures than there is at any time of the year, for by this time the silage and hay aftermaths are making their contribution.

PREPARATION OF LAND

Turning now to seedbeds, it is absolutely essential to create a really fine bed, which should also be a firm one. A fine seedbed promotes both an even distribution of seed and a more uniform depth of seeding. To ensure good germination and speedy establishment grass and clover seed should be sown at a depth of 12–20 mm. With seed at this level, seedling mortalities will be at a minimum, for the roots will get down to moisture even in dry conditions. Seeds which are on the surface, or which are barely covered will germinate if there is sufficient moisture in the soil, but they are very vulnerable should dry conditions follow germination—a circumstance not uncommon in April and May.

The certain placement of seed at the 12–20 mm level necessitates drilling rather than broadcasting, unless one adopts the safeguard of a heavy seeding rate to ensure that sufficient seed is at the optimum depth. The relatively new close-spaced drill is an ideal machine for drilling grass seeds. The costs are high but the recent development of contractor services for this purpose is very often the answer, and the saving in seed rate by using this service often more than compensates for the cost of the operation as well as ensuring in most circumstances a better germination and take. Many farmers who practise drilling use a corn drill in a double operation on diagonals to increase the amount of ground cover and to reduce inter-seedling competition. At its best this is only a half-satisfactory expedient. Spread of seed is limited and there is a waste of time, machine and labour, which are resources that should be conserved in our farming.

The case for drilling is stronger with spring seeding, because of the danger of early-summer drought, than it is with the later summer establishment which, except perhaps in years of protracted summer drought, is usually succeeded by favourable soil moisture conditions.

When a grass seed drill is not available the last cultural operation before broadcasting should be with the Cambridge roller, and preferably with one of sufficient weight to leave deep grooves to gather the seed and secure optimum coverage. A light harrowing will follow to cover the seed, and finally a further heavy rolling will complete the seeding process. It is vital to get adequate consolidation at this point. Too often one sees the failure to consolidate properly reflected by the vigour of establishment where the tractor wheels have made their impression, and by indifferent establishment elsewhere in the field. This is particularly noticeable on headlands which have been adequately consolidated.

An extremely useful roller to effect the sort of consolidation needed is that illustrated by Plate 10a. It is home-made, consisting of three 900 mm diameter concrete pipes which are filled with concrete and are fitted with stub axles. Each roller of the gang of three is only 750 mm wide, so that it follows the contours and there is a minimum of bridging if the ground is a little uneven. The whole unit weighs 3 tonne and you can certainly see where it has been after the operation. It is especially valuable for summer reseeds, because not only does it shatter surface clods, but it gives a really firm finish—a vital factor in securing success for late-season establishment. It has a special value, too, on soils which carry flints or stones, because they are pressed down out of harm's way. Machinery manufacturers have also produced very useful plain rollers for this purpose either concrete or water ballasted. An added advantage of water ballast is that the weight can be varied to suit conditions.

UNDERSOWING OF PASTURES

There are two further problems to be discussed for the benefit of those who still elect to establish their pastures under a cereal crop. The first is the timing of seeding, and the second is the choice of cereal.

Our preference is to get grass seeds in as soon as possible after the corn has been drilled. Apart from any other consideration, the earlier that seeding is effected the greater are the chances of sound establishment as a safeguard against early summer drought. Secondly, especially if the land is in good heart, or if normal fertilising is adopted, seeding before germination of the cereal will minimise the effects of competition. Seeding into a well-established cereal crop will be successful only under favourable conditions, and these one cannot expect every year.

Farmers' preferences for a companion cereal vary over the length and breadth of the country. In the south, the preference is for barley, because this is so often the last corn crop in the rotation. In addition it has the advantage of producing less straw to smother the seeds than either spring wheat or oats. Oats is a much more popular companion cereal in the north, primarily because it is often the last cereal in the rotation.

This consideration apart, barley is much to be preferred over spring oats. It has less straw, and it matures earlier, so that there is a better prospect of getting the field cleared in time to allow for a maximum of autumn growth. Spring wheat on the other hand, is invariably the last cereal to be harvested, although modern varieties have excellent standing qualities.

COMPANION PLANTS IN DIRECT ESTABLISHMENTS

Turning now to direct reseeding, it is generally advantageous to sow pasture seeds with a companion grazing nurse crop, which may be a cereal, or a light seeding of rape or soft turnips, or a combination of both. Our preference is for a cereal rather than rape, except perhaps on upland reseeds. Under favourable conditions the rape may grow too strongly even at low seed rates of 1–2 kg per ha and because grazing must be delayed for the sake of the stock until the rape is "ripe" there is a danger of smothering seeds.

Oats and wheat, being the more palatable, are preferable to barley if a cereal is used as a cover but usually the choice is determined by what is available. In most years there is some surplus treated cereal seed and the use of it as a grazing nurse crop is a convenient and gainful means of disposal. A seeding of 70–90 kg of cereal per ha is ample for this purpose.

There are two major reasons for including a grazing nurse crop when direct establishment is undertaken. First it results in a greater production of herbage, and secondly it is helpful in controlling competition from annual weeds. Grass, as well as clover seeds, are slow in making a showing in the spring, and especially is this true with slow-establishing species like timothy and meadow fescue. The cereal will make good growth within a month from seeding, to provide feed for stock at a time when weeds are in an early seedling stage and are relatively palatable.

Apart from grazing, trampling also has a deterrent effect upon weeds at this stage, and weed control apart, one cannot over-stress the importance of getting stock on to pasture as quickly as possible. Most Kent farmers believe that stock, preferably sheep, should be put on a pasture as soon as there is a reasonable blush of green showing. This belief stems from a recognition of the importance both of consolidation and of the value of fertility from excrements. Stapledon always stressed this feature of the golden hoof in pasture establishment.

When stocking pastures at this stage the best course is to put a lot of stock on for a short time, thus effecting a quick grazing, and then to rest the pasture until grazing is again justified. Above all, it is important to keep on top of the pasture in these early stages to ensure that there is adequate light for the development of clovers.

FERTILITY REQUIREMENTS

Mention of fertility leads us to a consideration of fertiliser requirements in grass sward establishment. Here one cannot generalise, but surely one cannot expect a fully-productive pasture if the land is acid, or if there is a deficiency of phosphate and potash. Where these are necessary, they should be worked into the soil immediately before seeding so that there is plant food within the root range of the developing seedlings.

Opinions differ as to the amount of nitrogenous fertiliser to be used in the seedbed. Certainly some added nitrogen is usually essential, especially if the reseed follows a cereal or Italian ryegrass, when soil nitrogen will be depleted. A total of 35–50 kg nitrogen per ha will be adequate to get the grasses moving. Even at this moderate level, care must be taken to ensure that clovers are not swamped by the growth of grass, by appropriate hard grazing.

When pastures are undersown in cereals intended for harvest it may be advisable, for the sake of the seeds, to broadcast the fertiliser rather than apply it to the cereal with the combine drill. If this is done, it is important to remember that broadcasting of fertiliser is only about half as effective as placement so far as the cereal is concerned. In other words, 500 kg of complete fertiliser which is broadcast has about the same effect as 250 kg down the spout. The combining of fertiliser may be preferred on land where there is a danger of the seeds "growing up through the band", but this is not a likely happening if the cereal is sown early at a normal seed rate.

MANAGEMENT AFTER ESTABLISHMENT

Management of undersown seeds after harvest should be directed towards the building up of root reserves before onset of winter. Avoidance of continuous grazing is preferred and it is wise to stock heavily for short periods alternated by longer periods of rest. Usually two grazings, with the second after the onset of hard frost, will be ample for these maiden seeds.

In their first spring it is preferable to use maiden seeds for grazing rather than for conservation. Here again it is a question of giving the clover a chance to develop, and to ensure that there is a build-up of fertility from stock droppings. This advice runs counter to usual practice in the north of Britain, where it is common to include some Italian ryegrass and red clover in the ley mixture in order to take a heavy hay crop in the first harvest year. On balance, it is better farming practice to graze a ley in its first season rather than take hay from it. If conservation is essential, then it is preferable to take an early cut of silage, if the pasture is to be fully productive.

ARABLE SILAGE AS A NURSE CROP

Finally, a word about using arable silage as a nurse crop for pasture establishment. The system is a useful one in that cover is removed relatively early in the growing season and there is enough time for the pasture plants to tiller out and make a sole. Our biggest objection to the method is based upon a dislike of arable silage as a feedingstuff.

Even when made under ideal conditions, there is not much milk or meat in it compared with pasture silage. Made under bad conditions, where there is soiling of the herbage, there is a considerable loss of palatability, and a depression of digestibility. Usually arable silage can be regarded as little better than a maintenance food, and its greatest virtue is that of a heavy crop. This means that, when it is removed, there is a severe depletion of fertility, and usually it will be necessary to apply about 400 kg of complete fertiliser to make good this loss and give a boost to pasture establishment.

CHAPTER XIII

MANURING OF GRASSLAND

ALREADY, in fairly general terms, emphasis has been given to the necessity for adequate soil fertility if the more productive species are to persist and give of their best. It now remains to give a more precise account of fertiliser practices. Before this can be done, however, it is necessary to debate the general economics of applying fertilisers to grassland.

Unfortunately it is impossible to give a satisfactory answer to someone who asks for advice on fertiliser practice without a very intimate knowledge, not only of the farm, but of the farmer's efficiency in using the grass he grows. Here we are up against the problem that, unlike a cash crop, grass does not generally become money until it is turned into milk or meat.

That indefinable personal factor which determines quality of management, together with variations in climate and soil, makes it impossible to give advice with any precision. At the top end of the scale one can confidently recommend farmers to spend as much as £90 per ha on fertilisers, because they are capable of getting an adequate return on their investment. Conversely, there are those for whom the safest policy is to use some lime or slag whenever soil analysis suggest a serious deficiency. One generalisation, however, is safe. For reasons already outlined, dairy pastures will almost always support higher fertiliser inputs than those grazed by sheep or beef cattle.

There is, however, a complication in the economics of fertilising dairy pastures in the shape of levels of concentrate usage. A dairy farmer is able to share his returns with both the provender and fertiliser trades and make a good living, up to a certain point. Beyond this he will lose out unless his stock and his stockmanship are of such a standard that he is able to get very high yields. If he is getting only average yields, both per cow and per ha, there is a very real danger that food costs per litre will be prohibitively high.

The implication is that a farmer who spends a lot of money on fertilising his grassland must intensify his stocking rate to ensure full utilisation, and this may mean lower yields per cow. At the same time, he must make every effort to ensure that the grass he offers his herd, whether it be grazing or in a conserved form, is of the highest quality. This latter provision will make for economies in the use of concentrates, either by feeding less per cow or by reducing the level of protein in the concentrate mixture. Whenever one examines the accounts of those dairy farmers who are making high profits out of intensive grassland farming, invariably they have a much higher concentration of stock and a much lower expenditure per litre on concentrates than are found on the average farm.

GENERAL PRINCIPLES

Regarding the type of farming, the starting point in designing a fertilising programme is an up-to-date soil analysis which gives details of the phosphate, potash and lime status of the farm and it is wise to have the soils of a farm re-tested at least every 4–5 years. Even farmers who feel that they have been reasonably generous in their manuring, may find no improvement, and sometimes a decline, in the fertility status of some fields, despite their generosity. Especially is this true on farms practising alternate husbandry.

On purely grassland farms there can be considerable drift of fertility, particularly in respect of potash, from one field to another. This comes about partly through the use of night and day fields, because there is more excretion by night than by day relative to the amount of food eaten in the two periods. Another factor is the repeated use of fields for conservation, with the conserved food being fed on another part of the farm.

The importance of soil analysis does not rest alone on indications of deficiency. It is equally important that a farmer should know where there is sufficiency, for there is no point in gilding the lily. The present-day popularity of compound fertilisers makes this point a particularly important one. Time and again one comes across farmers using a complete fertiliser more or less as a blunderbuss mixture in the belief that they are catering for everything. If the soil already has sufficient phosphate and potash they will get a response, but this will come from the nitrogen in the mixture. There is no point in making fertilising unnecessarily costly, for this response will be obtained more cheaply by applying a straight nitrogen dressing.

COMPOUNDS OR STRAIGHTS

This raises the broad question of the relative costs of straight fertilisers and proprietary compounds. The latter are very convenient, and there is now a very wide range of formulations to suit a wide variety of needs. It is almost impossible to make up and apply a fertiliser mixture more cheaply than a corresponding granulated compound, unless one is able to organise mixing as a wet day job. But if mixing delays application it will probably be cheaper to apply a compound.

One especially important consideration here is that a granulated fertiliser is ideal for the modern type of spinner fertiliser distributor, whereas powdery or crystalline mixtures are not. The former are evenly distributed, whereas with the latter, dressings are often very streaky. The spinner topdresser is cheap, it has a low maintenance cost, and a very high output.

Of course, it is not essential to mix fertilisers before application, because sometimes it is sound policy to apply straight artificials separately. The best example of this is to be found in the application of basic slag. This used to be one of the less attractive jobs on a farm, but nowadays, in most parts of Britain, it is possible to get a contractor handling bulk slag to apply a dressing which will suffice for three years or more.

Once this need has been satisfied and, assuming that there are ample potash reserves in the soil, further manuring can be limited to straight nitrogen applications at appropriate times of the year. If potash is seriously deficient this may necessitate a dressing with straight muriate of potash. Alternatively, if the potash deficiency is only moderate and fertilising also involves the use of nitrogen, then a nitrogen-potash compound can be applied. Where nitrogen is not required, the use of such a compound would be a needlessly expensive way of making good potash deficiency.

It does not seem wise to apply potash at heavy rates in the early spring, in view of a considerable suspicion that a "luxury" uptake of potash by quick-growing grass during the early flush period may be associated with grass tetany. In the present state of knowledge it is probably safer to apply any heavy corrective dressing in the middle of the growing season when clover is making its maximum growth, or else to use split dressings. There is an urgent need for more information on this point, because we cannot afford not to apply potash where it is a limiting factor to pasture growth.

This is more likely to be the case on light soils, such as the chalks and gravels, especially if a high nitrogen policy is combined with a heavy conservation programme. Twenty tonnes of leafy grass—the yield of a reasonably good crop of silage—

contain the equivalent of 125 kg of muriate of potash, so it will be seen that frequent cutting of a pasture for conservation can quickly deplete available potash reserves. Under these conditions, a policy of a little and often is probably sound. Hay on many dairy farms where the two-sward system of grassland management has been adopted and where the total nitrogen used often exceeds 350 kg per ha then a totally different approach to potash manuring must be adopted. On the grazing block some 70 kg of K_2O equivalent per ha is normally sufficient, while on the cutting block four times this K_2O equivalent must be applied; applying muriate of potash is cheaper than applying potash in a compound and it has the further advantage that it contains essential trace elements for plant growth. It is hoped that this product will be available for a long time. If phosphate is also deficient, then a complete compound fertiliser may be used, but basic slag applied at 2–3 year intervals to safeguard phosphate status is to be preferred on the score of cheapness and is normally sufficient for most circumstances.

TIMING OF APPLICATIONS

The timing of phosphate applications does not appear to be particularly important under British conditions. So far as slag is concerned, the most important consideration is to apply it when there is the least likelihood of doing physical damage to soils with heavy equipment, for instance during a period of hard frost or of summer drought. The same applies, of course, in applying lime.

One important consideration with slag is to avoid dressing a pasture to be grazed in the near future, for there is a very real danger of harm to stock if they eat herbage covered with slag. This danger generally disappears after a good shower of rain.

Timing of nitrogen dressings is a much more critical matter, especially in respect of applications for early-bite grazing. Here the advice is to stagger the nitrogen top-dressing programme, partly to ensure a succession of growth to meet grazing needs, but mainly to reduce losses by leaching during unfavourable weather conditions in the early spring.

When it comes to back-end nitrogenous dressings, the later one goes into September the less likelihood there is of a worthwhile response, although farmers in the more favoured districts have more latitude in this respect. One is entitled to take a risk, however, if grass supplies are short as a result of summer drought, as they were in the autumn of 1976. In this year, applied nitrogen did not show any appreciable effect till early October, when there was sufficient soil moisture. The ground was then still warm enough for nitrogen to give a worthwhile response, and this was particularly true of Italian ryegrass stands.

The colder and wetter the summer, the more important it is to get nitrogen on early for the autumn flush. In the North East, we had greater confidence in making this final dressing of nitrogen in August, but it still appears to be worthwhile to apply it to undersown Italian as the stubbles are cleared during the first ten days of September.

NITROGEN FERTILISING PRACTICE

Ammonium nitrate in the form of prills, due to its cheaper cost per unit than other forms, has taken over a very large proportion of the market. It must be remembered that like sulphate of ammonia it has no soil neutralising power and with continuous heavy use can induce soil acidity which must be corrected by the use of lime.

The whole question of the conflict between bag and clover nitrogen was debated in an earlier chapter, and it was pointed out that an intermediate levels of dressing there was a likelihood of falling between two stools. This is illustrated in the following table based on results obtained by Professor Linehan and his colleagues in Northern Ireland.

Average Annual Yields of Dry Matter from a Reseeded Pasture Receiving Adequate Lime, Phosphate and Potash.

(*kg per hectare*)

Applied Nitrogen	*Grass*	*Clover*	*Total*	*Response kg nitrogen*
0 nitro-chalk	4769	3891	8660	—
250 ,,	5899	2912	8811	1.0
500 ,,	6777	2385	9162	1.0
1000 ,,	8911	1004	9915	1.25
1500 ,,	10,542	251	10,793	1.42
2500 ,,	12,801	126	12,937	1.70

At levels of 250, 500 and even 1000 kg of nitro chalk per ha the response per kg in terms of dry matter was much smaller than it was at 1500 or 2500 kg. This is explained by the appreciable decline in the clover fraction which occurred even when only 250 kg of nitro chalk were applied annually. It seems that the applied nitro chalk was doing little more at the 250 and 500 kg level than make good the reduction of clover nitrogen it effected. It is noteworthy, however, that nitro chalk applied at a rate as high as 2500 kg per hectare was 70 per cent more effective per kg than it was at 500 kg per ha.

One sees in these figures the danger of adopting a compromise between the two extremes of fertiliser practice over a whole farm. But there is also a justification for applying 140–180 kg of nitro chalk in 5–6 dressings over the year, if the additional food is fully utilised by a remunerative process such as milk production.

In adopting a heavy nitrogen programme, there is no point in applying this fertiliser regardless of the availability of grass, whether it is for stock or for conservation. There must be a very high measure of opportunism in using nitrogenous fertilisers. If there is a pressure on available grass, then immediate applications will be justified, at least on part of the grassland, but if grass supplies are adequate then the fertiliser is best kept in the store. This may also be true during drought periods when water, not nitrogen, is limiting growth. Fertiliser nitrogen must be a sensibly used tool and not an addiction in intensive grassland management, although the problem is to predict climatic conditions in good time and apply the appropriate quantity to maintain a steady supply for stock.

LIMING

Finally a word about liming. It is generally accepted that a pH of 6.5 is the optimum for grassland. At the same time, it is not advisable to overlime. Moving deliberately above a pH of 7 will almost certainly reduce the availability of phosphate and it may induce certain minor element deficiencies in subsequent tillage crops. This brings us back to the importance of soil analyses. It is a wasteful, and even dangerous philosophy, to work on the principle that because a little of something does good, a lot of it will do a lot of good. Certainly this is not always true with lime.

CHAPTER XIV

IMPROVEMENT OF PERMANENT PASTURES AND ROUGH GRAZINGS

THE virtues of good permanent pastures have been extolled, but this must in no sense be interpreted as an open testimonial, applicable to all permanent pastures. To the contrary, for the overwhelming proportion of them leave a great deal to be desired. In many cases the circumstances are such that they should be ploughed at the first convenient opportunity, so that a fresh start may be made to provide a basis of more desirable species, or else to create a surface that permits mowing as part of their management. In other cases there will be a sufficient foundation on which to build, using surface improvement techniques only.

Sometimes some surface improvement must be undertaken before it is advisable to plough. The most important pasture group falling into this category consists mainly of upland grazings, where there is a layer of peat with a cover of low-producing acid-tolerant grasses, rushes, and possibly heather. If this layer is ploughed in it may remain undecomposed at ploughing depth for many years to impede drainage, prevent root penetration, and create ideal conditions for rush invasion. The correct approach to reclamation on such land is to break down this top layer by surface improvement methods before ploughing is attempted, if ploughing has to be used at all.

Drainage usually has to be undertaken as a first operation because normally such soils are poorly drained. This will be followed by the application of lime and phosphates and, above all, by the concentration of stock to eat out dominant species and to effect hoof cultivation which assists in breaking the surface mat. More breeding cows on the Welsh hills for example could result in a rapid acceleration of improvement of swards based on *Molinia* and *Nardus* with some fescue present.

USE OF STOCK

Here one is thinking of stock as tools in pasture management. No-where has this function been illustrated to better effect in recent years than in New Zealand, where "mob-stocking" following aerial top-dressing with phosphate has transformed inferior store stock pastures on steep unploughable land to a point where they are now capable of producing fat lambs. Mob-stocking implies the concentration of a large number of stock on a limited area for relatively short periods of time.

Short periods are stressed because stock which are made to work for a living in this way cannot be punished too severely, and it will generally be necessary to give them some relief from their labours. This necessitates subdivision of the area to be reclaimed, at least down to the field unit size which will ultimately be adopted on the farm.

Ewes or single-suckling cows after weaning can be used for this purpose, but their availability is limited to a relatively short period. The ideal time to hit *Molinia,* a characteristic grass of such areas, is in the early summer period when it is starting to make growth and has some semblance of palatability. In-calf heifers, up to two or three months before they are due to calve, are admirable for this work, but one has to use considerable judgment in handling stock in this way, so that harm is not done to them, and one must be prepared to provide good feed later to compensate for any set-back in condition. The best approach is to have three or four enclosures that are being improved at the one time, so that some form of rotational grazing can be practised. This will be good for the stock and also for the work in hand.

SURFACE SEEDING

Provided land is reasonably dry and there is some natural shelter, good work will be done during the late autumn and early winter by cattle receiving supplementary feeding which includes fairly mature hay, because this results in a considerable measure of seeding. Another device for introducing clover to such areas is to lay up an already improved area of the farm in the summer to allow clover seed heads to mature. If these are eaten by sheep which are shifted from this grazing to the areas under improvement a number of times, there is a very appreciable seeding of clover which will become effective if phosphate and lime have been applied in sufficient quantities to allow the clover to establish.

Then, of course, there is the straightforward approach of broadcasting grass and clover seed on a surface that has been cleared by burning, hard grazing or even light rotovation, and using stock to tread in the seed. Under some circumstances, harrowing in of seed can be very effective. Surface seeding in this way has been effectively practised in parts of Scotland on thin soils where outcrops or boulders prevent normal cultivation. Initial progress by such methods may seem slow, but usually there is an acceleration once the process gets under way and soil conditions start to improve.

Someone with plenty of capital who is on cultivatable land will usually prefer speedier methods, such as pioneer cropping followed by direct reseeding. But the farmer with limited capital, both for land improvement and for stocking pastures once they have been improved, will find that this gradual approach will suit him very well. He can raise his sheep and cattle numbers by a process of natural increase rather than by purchase, and in this way he maintains his farm in equilibrium.

Too often one encounters examples where men have used most of their available capital in large-scale improvement schemes and have insufficient livestock to exploit these improvements. If improved pastures are not stocked adequately, they will soon deteriorate as a consequence of invasion by the less desirable species that are encouraged by undergrazing.

When dealing with marginal land with only limited capital it is usually advisable to work to a long-term plan, with ploughing and reseeding playing only a limited part on those areas which are ready for this more expensive method of land rehabilitation.

The more recently developed technique of chemical ploughing, using one of the new materials which destroy grass is potentially a valuable aid to the improvement of such land. Once the herbage has been killed a match can be put to it on a dry day without too much wind, and trash will burn to leave a reasonably clear surface for scratching with harrows and over-sowing with grass and clover seeds or by the use of the numerous sod-seeding machines now available provided the land is tractable. Once established, however, good management is particularly essential to prevent deterioration by invasion of rushes. It must be stressed, however, that marginal land farming cannot afford expensive inputs, because usually the occupier has to make his living at the same time as he improves his farm.

LOWLAND PASTURES

When one comes to more tractable land at lower altitudes, problems of permanent pasture improvement are not nearly so difficult. In fact, on a great deal of this land, though grasses like *Agrostis* may dominate, there are often already appreciable proportions of the more desirable species such as ryegrass, timothy and white clover which can be encouraged if the fertility level of the soil is raised and intensive grazing is practised.

Anyone who saw the Treefield experiment at Cockle Park before it was ploughed out will appreciate the importance of these factors in transforming derelict permanent grass. Plot 6, the control, continued for sixty years as a low producing association of bent grasses and fine-leaved fescues with scarcely a trace of wild white clover. The adjacent plots, which received 625 kg of slag per ha at three-yearly intervals to make good the principal deficiency which was phosphate, had approximately five times the productivity of the control. The process of improvement is a cumulative one, for with the increase in clover due to slagging, increased stocking is possible. This, in turn, through the larger returns of excrements, raised the fertility status to the point where productive species like perennial ryegrass are dominant in the sward.

It is not enough to apply fertilisers. One must also create the other biotic conditions which will favour the spread of productive species and which will discourage weeds—for example, drainage where this is necessary. It is also very important to let light into the sward, for both ryegrass and white clover require light as well as fertility if they are to tiller freely. Nowhere has the importance of this light factor been better illustrated than on some naturally fertile but under-drained land in the Romney Marsh district. The first step was the very necessary one of lowering the water table by cleaning out dykes and by effecting general improvements in drainage. The second was the repeated use of the gang mower in association with intensive grazing. Formerly the pasture contained a large amount of rough grass interspersed with horsetail, rushes and thistles, and it provided nothing more than some summer keep for store stock. Today it is a dense mat, predominantly ryegrass and white clover, and it can be classed among the most productive pastures in this country, regularly producing well over 500 kg of liveweight increase per ha.

Not many farmers have gang mowers, or indeed surfaces which will carry them, but they have other tools which will do the same sort of job—namely their grazing animals, mowing machines and forage harvesters. One way of dealing with a rough type of permanent pasture which carried dead growth is to give it a fairly good application of nitrogen, say 60 kg N per ha, in addition to its basic fertiliser needs, and to take a silage crop which is cut before the crop gets yellow in the bottom. There will be clean, fresh growth in the aftermath, and because nearly all the rough growth has been removed there will be a chance for the clover to run during the summer, provided the aftermath is not permitted to grow too strongly. Here again, a programme of heavy stocking for short periods followed by rests is ideal to promote this advantage for white clover.

CONTROL OF WEEDS

Sometimes the preliminary step must be taken of reducing weed population before such fertilising, cutting and grazing management can be exploited to full advantage. The most serious weeds of old grassland are daisies, buttercups, thistles (especially creeping thistles), rushes and docks.

Improvements in drainage and fertility levels are in themselves methods of reducing the incidence of some of these weeds, especially buttercups, and rushes. One of the most spectacular features of the Hanging Leaves experiment at Cockle Park was the virtual absence of rushes on the drained area of the field. On the undrained portion there were relatively few rushes where slag was applied, but where slag was omitted rushes were the dominant vegetation.

Today hormone weed killers can be utilised to advantage in

controlling many weeds in grassland and considerable success has been achieved in handling rushes in this way. MCPA is commonly used, although it is likely that this herbicide will impair clover growth if this species is present. Less harm will be done if the chemical is applied fairly early in the spring, before clover starts to grow actively. However, if clover is already very deficient, it may be advisable to ignore its welfare in the first instance and make a job of clearing the weeds, depending on subsequent surface seeding with clover to make good this deficiency.

Rushes, of course, will not stand repeated mowing, but on much land carrying a dense plant of rushes, mowing is out of the question until they have been severely reduced by spraying and drainage. Even then it may be necessary to plough and level, in order to get a reasonable mowing surface.

Mowing is also one of the most effective methods of dealing with thistles, whether they be in permanent pastures or leys. Mowing is less effective if it is just a matter of topping, because the rosette form taken by thistles under grazing conditions helps them to survive cutting. If, however, the pasture is allowed to grow to the silage stage twice in the one season and the thistles are cut when they are in full growth, there will be very few remaining in the following year.

Docks are a more difficult problem, especially on dairy pastures where fertility is high, but they are much more of a menace in ley farming than in well-managed permanent pasture. The aim must be to prevent them setting seed, and here silage making is infinitely preferable to hay making, especially when hay is cut at a very mature stage. Mature hay, carrying ripe dock seeds, is an abomination, because it is probably the main agent in spreading the dock nuisance.

Docks were becoming an increasing weed menace up to the introduction of specialist dock sprays a few years ago which, although expensive, are reasonably effective. Hard grazing with sheep is almost as effective as any other method of dealing with docks. It is also an excellent way of controlling ragwort because sheep will eat out the crown of this noxious weed to prevent seeding and further spring.

SOD SEEDING

Sod seeding is now a practicable proposition on permanent pastures on the better soils where there are no problems of destroying a surface mat or rough vegetation. In Australia, New Zealand and North America, sod seeding is well past the experimental stage and is now part of farm practice. A common New Zealand method is to fit special cutting tips to a drill which has hoe-type coulters. These rip the turf and expose sufficient soil to allow the seeds to strike. Where there are no stones, a disc coulter drill can be very useful, if it is used when the soil is fairly moist. This not only results in a more effective cut, but it gives the seeds a better chance of germinating. As mentioned earlier some types are based on modified strengthened Suffolk type coulter drills or disc drills, and others are based on the rotary cultivator.

If conditions allow, rolling should follow the seeding. If this is not possible, then mob-stocking should be substituted, not just immediately after seeding, but at frequent intervals thereafter to reduce competition and over-shadowing by the originally established species.

There are two periods of the year when sod seeding can be attempted—either in the spring or in the late summer, i.e. the July-August period in Britain. At either time the pasture should be eaten bare, even to the point of being "poached red" as our Irish friends would say. Spring seeding will possibly give a better prospect of clover establishment, but here again it is a question of ensuring that there is no smothering of seedlings, by appropriate grazing. Perennial ryegrass, which is very quick to establish, especially with summer seedings when the ground is warm, is the most certain grass to sow. In New Zealand there have been impressive results with *Grasslands Manawa* ryegrass, and oversowing has now become a standard method of introducing this valuable out-of-season producer into a permanent grass system without recourse to the plough.

Seed rates of 12 kg per ha of ryegrass and 2 kg of white clover per ha appear to be ample when the seed is drilled in, but somewhat higher seed rates will be necessary, say 18 kg in all, if broadcasting following some discing and harrowing is done on a bared pasture. We used this last method quite successfully at Cockle Park in a wet July-August period when there was ample moisture to promote germination and seedling survival. A double cut with the discs exposed enough soil to get a take, and this was safeguarded by heavy rolling immediately after seeding. A modified system used by the junior author at the Welsh Agricultural College has been rotary cultivation at a depth of 5–7 cm on old grassland using the special tines provided by the manufacturers to break the pan so often formed by rotary cultivators. This has resulted in a system where stiff yellow clay some 10–12 cm below the surface is not brought up and is often helpful in a dry autumn as there is less moisture loss than there is with ploughing.

IMPORTANCE OF MANAGEMENT

Once a permanent pasture has been brought into reasonable shape, it is important that it is not allowed to backslide to its former state of semi-productivity. Avoid over-grazing in the spring when there is a danger of poaching, for invariably this will encourage weeds like buttercups and daisies, especially on strong land. Later in the season it is equally important to avoid the twin evil of undergrazing, for this will give rise to neglected clumps of rough herbage which will weaken the clover and encourage poorer species like *Poa trivialis* and Yorkshire fog.

Repeated cutting for hay, especially at an advanced stage of growth, has a similar deleterious effect. Apart from the danger of depleting fertility, tall growing species come into an ascendancy at the expense of ryegrass and white clover, which must be regarded as a climax association in a permanent pasture. Occasional cutting, however, especially at the early silage stage, is sound practice, for it removes the clumps that will be neglected by stock before they cause harm.

Above all, there should be no fear of eating a permanent pasture bare in the late summer immediately prior to the onset of the autumn flush. Not only will this give a better quality of back-end grazing, but it will help pasture composition. A permanent pasture should not be put up for a strong growth of foggage year after year, however, because this will lead to cocksfoot and Yorkshire fog becoming dominant in the sward.

Finally, the precaution of obtaining soil analyses at regular intervals is stressed, so that the all-important consideration of plant nutrient status is safeguarded.

CHAPTER XV

CONSERVATION OF GRASS

ANY method of conservation of grass—as hay, silage and to a lesser extent as dried grass—entails waste, in that the final product fed to stock invariably has a lower feeding value than the grass when standing in the field. A striking example of this is silage. Although cut in mid-May, when one would expect a dairy cow to maintain herself and produce 18–24 litres of milk per day, it ends up as a product that even under the best conditions of maintaining a cow with sufficient energy for 9 litres of milk and enough protein for 18 litres. Under poor conditions, with an indifferent fermentation and surface waste, this difference between the raw material and the finished product may well be more. Haymaking can show greater losses still, up to a point of total loss when a crop rain-soaked for some 3-4 weeks may have to be disposed of by setting fire to it.

It is interesting to note that the senior author in the 1967 edition of this book remarked that "Though grass drying is the safest method of conserving grass in respect of quality of the finished product, as well as reduction of conservation losses it is much too expensive for the ordinary farmer". Current cereal prices and particularly world protein prices may well see a renewed interest in the production of dried grass for stock feed, particularly as a source of protein, always providing fuel costs do not rise disproportionately.

SWING BACK TO SILAGE-MAKING

Since 1940, when the drive for more silage got under way, farmers have been adept in finding good reasons why they should not make the stuff. Undoubtedly there was much point to their argument that too much heavy labour was involved in getting material in and out of the silo. The validity of this argument has been lessened first of all with the introduction of buckrakes and, later, of relatively cheap and very efficient forage harvesters. The development of self-feeding or mechanised easy-feeding of silage has been a further inducement to popularise silage-making as an attractive alternative to hay-making in this unreliable climate of ours and has resulted in a very large swing towards silage-making in recent years.

A second shortcoming of the silage process weighing heavily in farmers' minds was an uncertainty that they could regularly produce a first-class product. In a way this was the fault of those who were concerned in giving advice on silage-making in the early years of the war. They made the process sound much too complicated—and did not help matters either—by suggesting that silage-making could be undertaken in any weather. This is quite erroneous, for if good silage is to be made the herbage must be without free moisture. In fact, there are strong arguments for some wilting of herbage, for as results given later will substantiate, a high dry-matter content is very important in determining the feeding value of silage.

A further factor was the widely current view that there was some magic in the process that made a silk purse out a a sow's ear. The senior author can remember, in 1947, a farmer of some standing who had written an authoritative article on silage-making showing him a clamp made with the first cut from a new stand of lucerne which was dominated by fat hen. He was hurt by the suggestion that it would have been less bother to have put this rubbish directly on to the muck midden.

COMPARATIVE VALUE OF SILAGE

The simple truth is that silage is never any better, and very often much poorer, in feeding value than the original material put into the silo. In fact, with herbage cut at the same stage of growth and made into silage or hay, the probability is that hay made without weather damage and a consequent loss of leaf and palatability will have the higher nutritional value. This is illustrated by the following American figures* derived from *ad lib* feeding of three kinds of conserved food from the same crop to 300 kg heifers:

Food	*Dry matter*	*Dry-matter intake per 100 kg body wt*	*Average daily liveweight gain*
Unwilted silage	17.9%	1.60 kg	0.04 kg
Wilted silage	25.0%	2.01 kg	0·58 kg
Hay	84.1%	2.33 kg	0.76 kg

These figures stress the importance of a high dry-matter content in influencing the intake of dry matter by cattle and, in turn, the rate of liveweight gain. Similar experiments have been undertaken in the feeding of dairy cows, and here again one gets corresponding differences in dry-matter intake, though less marked differences in levels of milk production. It almost seems that silage dry-matter is converted more efficiently into milk than hay dry-matter from the same crop, but this does not appear to apply to liveweight gains.

The case for silage as opposed to hay does not rest, therefore, on any superiority of silage as a feedingstuff, provided both are made from the same sort of material under ideal conditions. This last qualification is the rub in Britain, because it is so seldom that

* Everett, Lassiter, Huffman and Duncan, *Journal of Dairy Science,* 1958, p 720.

PLATE 13a

The production of weaners from single-suckled cows is not sufficiently intensive to provide a good living for any but large-scale farmers who have relatively cheap land.

PLATE 13b

The dairy herd coming up the race for afternoon milking on Mr Charles Platt's Woore Hall Farm, Woore, Salop.

ICI Photo.

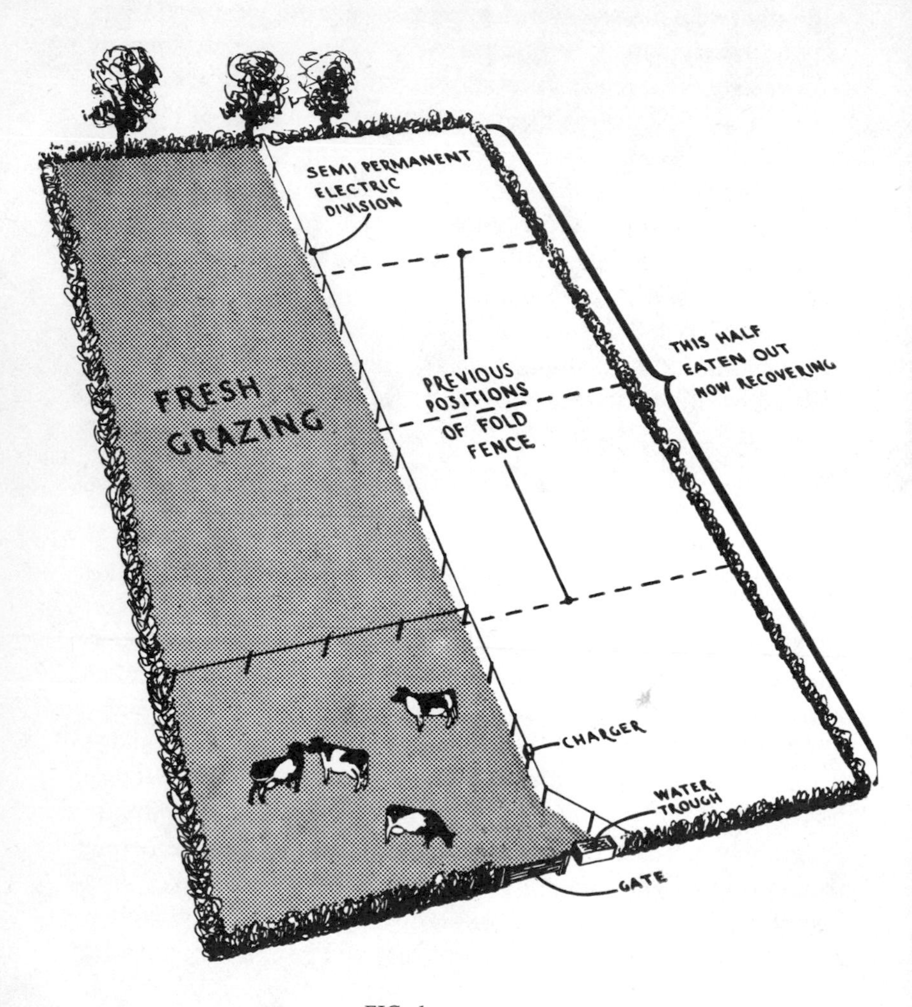

FIG. 1

Layout for fold grazing to protect recovery growth and to minimise labour in daily shifting of fold wires.

we have sufficient good weather to make really first-class hay. Although we usually have our best hay-making weather in late May and early June when pasture is at the appropriate stage of growth to make a really good product, the herbage is invariably so full of sap that drying is prolonged and the danger of weather damage is aggravated.

Quick hay-making methods and barn drying, which will be discussed later, may bring about greater certainty of saving early-cut hay in a reasonably good condition. For the present, however, the policy of the great majority of farmers is one of leaving hay to an advanced stage of growth, partly because this will mean quicker drying and partly because it will mean more bulk. In practice, therefore, the debate is on the relative merits of silage cut early in the season and hay cut at an advanced stage of growth.

Provided a farmer is equipped to make and feed silage economically, the arguments seem to be mainly in favour of silage if the forage is to have a production as well as a maintenance function. Most of these arguments have already been advanced—the higher digestibility, net energy and protein values of material cut at an early stage of growth, the leafier and more vigorous aftermaths that follow early cutting.

One further cogent farm management point in favour of silage is that every ha cut for silage in May and early June is a hectare less to cut for hay later in the season. A silage and hay programme spreads demands on labour and gives a comforting feeling that all eggs are not in the one conservation basket. In a bad year a farmer who does not secure the bulk of his grass crop as hay or silage by the beginning of July will have a very anxious time salvaging the wreckage that the worst sort of British summer can create.

The issue, of course, need not be one of silage versus hay. Very often in practice it is wise to make a lot of silage but some really good hay in addition. Very often our best hay-making weather comes early in the season, and if this is the case silage-making should be interrupted to make the most of the opportunity.

Repeated tedding, started immediately after mowing, or the use of a flail forage harvester as a hay-making tool reduces much of the risk associated with the early cutting of hay. It pays to be an opportunist in hay-making. The longer the season of operation and the lower the target, quantitatively, the greater are the chances of making really good hay.

One point is certain, however. Nothing will be gained if large quantities of over-mature hay are substituted for good silage. We had an experience at Cockle Park that bears out this point. The herd was on self-feed silage of good quality, which was supplemented with early-cut hay but which was not particularly palatable because it had suffered some weather damage. This was replaced by very mature timothy/meadow fescue hay cured in a barn drier. The cows ate this avidly in preference to the silage, and the immediate result was a substantial drop in milk yields. These recovered, and so did silage intake, when we changed back to the original quality of hay.

Where cows are tied by the neck and are rationed individually one is able to control the intakes of the various kinds of feed, but in a loose-housing system with *ad lib* feeding of hay and/or silage it is essential that both are of the highest quality. Silage enthusiasts will argue that it is simpler to rely on silage alone under these conditions, a view which is also maintained by the junior author. At Frondeg Farm with the spring-calving herd, their fit condition at calving and their subsequent level of production bears testimony of the suitability of good silage as the main constituent of the diet provided it is of good quality i.e. fermentation, digestibility and palatability.

PRECAUTIONS IN SILAGE-MAKING

Turning now to the silage-making, the important considerations are that it must be carried out cheaply and expeditiously, and a good product must be obtained. So far as the first object is concerned, the combination of a flail-type forage harvester with easy-hitch tipping trailers and a buckrake at the clamp is difficult to improve upon at the present time. Provided haulage distances are not excessively long, a team of three men with two tipping trailers and a buckrake to put herbage on to the clamp, and in the process to effect necessary consolidation, will handle 60–70 tonnes of herbage in a fair working day. This means, among other things, that a crop can be saved before it becomes too mature, and that one can generally afford to pick the right weather conditions.

This last consideration is all-important in securing the second object, namely good quality silage, for it is very unlikely that this will result if the material being ensiled is saturated with rain water. Not only is there a grave risk of an undesirable type of fermentation, but there is the added disadvantage that the silage will have a low dry-matter content, which will impair its feeding value.

The simple "in line" flail forage harvester is often ideal for making a small quantity but it can also be used, as on Mr Edwin Bushby's farm, for fairly large tonnages provided the haul is short; in fact he wilts the grass first and then transports the silage to the clamp without unhitching the "in line" harvester (see Chapter XX).

There is little doubt that a "double chop" machine makes a better silage, mainly due to laceration, release of plant juices, better consolidation and fermentation, and this silage is far easier for self-feeding with its shorter chop. Our experience with this machine, however, is that, unless the blades are kept correctly adjusted and sharp, there is a loss of "chop" as the season progresses, ending up with a product with no finer chop than a flail cut product, with consequent difficulties when self-feed silage.

The ultimate machine at the moment is of course the far more expensive full or "metered chop" machine with tine pick-up from a wilted swath or, in one case, a direct cut machine using an incorporated disc mower. Recent results from Bridgetts Experimental Husbandry Farm have shown that the same raw material cut by a full chop machine made better silage, with the better fermented product achieving a high liveweight gain in beef cattle. The cost of these machines, both for initial purchase and for maintainance and repair is high, but there is no doubt that their use is beneficial provided sufficient silage is made annually to justify their extra cost. The advantage of full chop silage with 4–5 cm chop is obvious in self-feed systems but unless precautions are taken it can prove a disadvantage in that the silage face is loose and easily pulled down by cattle, with consequent wastage.

ADVANTAGES OF WILTING

The importance of wilting is now accepted by most successful silage makers. Briefly the advantages are:

(1) More dry matter can be carted per load of a given cubic capacity, giving consequently a quicker rate of silage-making.

(2) Normally a better fermentation is assured with a silage of high dry-matter content.

(3) The animal is able to eat a higher level of dry matter per day as the silage is less bulky due to the lack of extraneous water. This higher level of intake is usually enhanced further by the better fermentation associated with wilted silage.

(4) There is far less and even no effluent from a clamp of wilted silage.

There are certain disadvantages of wilting. These are:

(1) Grass cut for wilting may deteriorate more quickly in bad weather than a crop left standing for direct cutting.

(2) An extra operation, cutting the grass, is needed. This is difficult in the smaller farm situation, but in good weather the extra operation is often more than compensated for by the increased rate of removal of the cut grass from the field.

(3) Care must be taken not to allow wilted silage to overheat in the clamp. This can be prevented by hard rolling or better still by temporarily sheeting the clamps at night, and on days that no grass is being carted, with a polythene sheet draped over the whole surface.

The tower silo, of course involves more expensive machinery and in this system it is absolutely essential to wilt particularly as unloaders and augers will not work with low dry-matter silage.

The general belief has been that the optimum temperature for silage-making is about 38°C, on the grounds that if one allows temperatures to rise to this level there is less risk of the undesirable butyric type of fermentation which has contributed so much to the unpopularity of silage, especially with farmers wives. There is, however, little experimental evidence to show that this level of temperature is in fact the right one. An increasing body of opinion, notably encouraged by the work of Professor Murdoch when he was at Shinfield, believes that a low temperature (25°C) is preferable, in that it results in lower nutrient losses and a better fermentation.

The principal reason for confused thinking on this point is that the ensiling of wet material generally results in a low-temperature fermentation, with the production of butyric acid. The butyric acid is, in fact, the result not of low temperatures but of excessive moisture. This is borne out by Murdoch's work, which shows that top-quality silage, resulting from a lactic acid fermentation, can be obtained from the cold method provided dry material is ensiled.

Another factor which probably influences the type of fermentation is the sugar content of herbage. This is most noticeable when comparing silage made in a long sunny spell with silage made in cool muggy weather. The former is better as not only is the grass drier, but contains a large amount of sugars as photosynthesis is taking place at this time at a much higher rate. These sugars are available for suitable fermentation to take place. Where one is dealing with a very young material or one with a high legume content, or in damp conditions (for example, a mainly lucerne mixture) a good if rather old-fashioned method is to add molasses at the rate of 6–10 kg per tonne of green material, to provide the necessary substrate for the proliferation of lactic-acid-producing bacteria. This result can also be obtained by incorporating modern additives when silage-making.

PRECAUTIONS TO TAKE

In brief, then, the precautions to take in making good silage are:

(a) The use of leafy material harvested before ear emergence and ensiled without free moisture.

(b) The addition of sugar in the form of molasses if there is a high legume content, or the use of additives which we will discuss later.

(c) Effective consolidation of the mass, whether it be in a clamp or a tower, to prevent a rise in temperature. If weather conditions or other circumstances interrupt silage-making, the clamp must be rolled heavily and covered with a polythene sheet to prevent any rise in temperature—this is especially important if the material has been wilted. The effect of the polythene sheet is that it traps carbon dioxide in the clamp, thus effectively slowing down respiration which is the process that releases heat into the mass.

Finally, once silage is made, rainwater must be prevented from penetrating the mass, and here lies the importance of making silage under a Dutch barn which gives a complete cover or in an effectively sealed open silo. Much of the harm done by rain comes when the clamp has been opened for feeding—half an inch of rain will penetrate deeply and the stock will be offered silage with a high moisture content. Dr. Dodsworth, in his trials at Aberdeen, showed that the deliberate adding of water to silage at the point of feeding greatly reduced the dry-matter intake by stock. A high dry-matter intake is of paramount importance if one is to get the best out of silage when it constitutes the main component of an animal's diet.

Another factor influencing intake of dry-matter is stage of cutting. At Cockle Park a trial concerned with feeding silage to sheep tested differences in intake between silages from the same field, with one made from herbage cut just prior to ear emergence and the other cut three weeks later. There was very little difference in moisture content or in the type of fermentation, yet the sheep on the young silage ate approximately twice as much as those on the mature silage. This result was to be expected, because rate of passage of material through the gut is positively affected by digestibility. This in turn has a direct effect on appetite.

The introduction of recent additives based on formic acid and a mixture of formaldehyde and sulphuric acid applied with very efficient drip applicators attached to forage harvesters has resulted in much reasonable silage being made under conditions that would normally have resulted in very poor silage. This is not the only use of these additives. When wilting is impossible, their addition helps to produce better fermentation and they are particularly useful in autumn silage making. We would go as far as to say that no autumn silage should be considered without additives as at this time grass is often damp, low in available sugars and without additives it results in a product of low acceptability by animals and low feeding value.

It is not suggested that additives should be used when making wilted silage in late May and early June in hot sunny weather, but they should be on hand in case one has to gather the crop in fairly damp muggy weather when wilting is not possible.

HAYMAKING

It has long been recognised that if really good hay is to be made, conventional haymaking methods are not satisfactory, especially in the higher rainfall areas. Ironically, however, the best hay used to be made in the districts with the more difficult haymaking conditions, simply because the farmers concerned took the necessary precautions to minimise damage either by piking or using tripods. This situation has not been maintained, however, with the increasing popularity of the pick-up baler and the steady rise in labour costs.

Though nobody is likely to question the quality advantage from making hay on tripods, the time and labour involved made it a very doubtful economic proposition. More and more in our farming we are being forced into a position where we have to discard an operation if it cannot be conveniently mechanised. Tripodding comes into this category, and so endeavours to safeguard the quality of the hay crop are now taking other forms, in particular "quick" haymaking methods and the in-barn or in-stack drying of herbage baled at a stage where normally it would be considered in need of a further day's drying.

Quick haymaking may take one of two forms. The first is the more conventional one of repeated use of the tedder immediately after cutting so that the swathe is completely exposed to drying agencies. If one leaves the swathe undisturbed until it is dry on top and then turns it, especially if this is done by a finger-wheel type of machine, there will be a sad core of green material in the centre of the swathe when the rest of it is fit for baling. The finger-wheel swathe turner is an excellent machine for the purposes for which it is intended, namely turning and rowing up, but it is not a tedder.

Much of the indifferent hay made by pick-up baling is attributable to the failure of farmers to purchase an inexpensive and highly efficient power take-off driven tedder which will really fluff-up the crop. Given low humidity and sunshine with some air movement, it is reasonable to expect that a comparatively heavy crop of 4,500–5,500 kg/ha can be made ready for baling within 72 hours from cutting, if the tedder is used to full advantage. The use of a bruiser-crimper has not become popular even though it speeded up drying time considerably by cracking and bruising stems.

The extreme in quick haymaking is the use of the flail mower instead of a mowing machine. We investigated this at Cockle Park, and our general conclusion was that it is possible to make excellent hay with the flail forage harvester provided full advantage is taken of a short break of fine weather. However, there was considerable loss of leaf in the process.

Where there was no weather damage the quality, as judged by protein content, was slightly superior to that of conventionally made hay, but if there was broken weather the chopped material deteriorated very quickly. The risk of weather damage is considerably lessened, however, because the drying period is appreciably reduced.

On several occasions under ideal conditions hay was cut at Cockle Park, baled and carted within a period of 36 hours. In other words, the time taken for the whole process was more than halved as compared with normal mowing and tedding. There were losses in yield, however, and these have amounted to about 12 per cent in a 5 tonne crop. Percentage losses were relatively higher in a light crop than they are in a heavy one, but there is less justification for using the method with light crops.

There can, however, be compensations for these mechanical losses. In one trial in 1958 we harvested the lacerated material in good conditions, but before the conventionally-treated herbage could be baled the weather broke and anything that was carted off the field was the sort of hay to use only as a last resort.

Anyone attempting quick haymaking methods with these fairly drastic treatments of the swathe is well advised to take good notice of the weather forecasts. If these promise fair for three days then he can feel reasonably secure, but repeated wetting of lacerated material will result in its rapid deterioration on the ground or in bales. It is a very wise precaution to cart the bales immediately after they are made, because they will not turn rain.

IN-BARN DRYING

The system known in Britain as in-barn drying, and in the United States as mow-drying, has some enthusiastic supporters in this country, and with considerable justification, too, because really first-class hay can be made by this method. The process follows conventional cutting with immediate and repeated tedding until the moisture content of the material is reduced to about 40 per cent, when it can be baled, provided the sun is shining and there is a fair measure of crispness in the herbage. Under good conditions this will be some 54 hours after cutting.

The bales may be stacked, cut edge down in a chamber, if walls are available, and the air distributed through a steel mesh false floor some 50 cm above ground level. More layers of bales are added periodically to the lower layers as they dry out. Another popular alternative is to have a series of ducts branching from a main duct running the length of the Dutch barn for distributing the air. Some heat may be provided by covering the tractor or engine so that in-coming air passes over the hot engine and picks up a considerable amount of heat in this way.

At Tanygraig Farm the junior author has been most fortunate in inheriting a roofed silo, built before the turn of the century with 7 m high walls. It was a relatively simple task to convert this into a very effective barn hay drier by putting on a false floor and a tractor driven fan. In the early summer of 1972 characterised by the very wet weather, comments were made frequently as very heavy bales of hay were placed in the chamber. These critics were very pleasantly surprised with the resultant product that came out to be fed to calves in the winter of 1972/73, enabling us to have a first-class product. Most of our neighbours had to wait for a few weeks for hay-making weather, and although their product smelled good, its coarseness due to the quality of the crop, resulted in a very low feeding value.

If no effort is made to save quality material, there appears to be little justification for barn drying and its attendant additional expenses, but regarded as a cheap half-way house to grass drying, especially on the smaller dairy farms in the high-rainfall districts or on lucerne-growing farms in the eastern counties, there is much to be said for the system. We certainly did not regret our decision to install a barn dryer at Cockle Park. Apart from other considerations, it has been a source of great comfort to those responsible for management of the farm during some particularly difficult hay-making seasons, as well as to the stock utilising its products. It was stressed earlier that a lot of good silage and some good hay was the best approach to pasture conservation, and there is no question but that in-barn drying is one certain method in achieving this end.

CHAPTER XVI

FAT LAMB FROM GRASS

THE main factors, apart from business acumen, in determining success in the production of fat lamb from grass are as follows:

(a) The fecundity and milking ability of breeding ewes
(b) The inherent growth rate of lambs
(c) The effective control of disease
(d) The quality and cost of the feed supply
(e) Flock depreciation
(f) Efficiency of labour use
(g) Flock management policies.

Obviously we cannot deal with all these factors in detail, for this would constitute a book in itself, and in any case some of these aspects are outside the scope of our discussion. They are mentioned because it is important to realise at the outset that fat lamb production, like milk production, depends for its success on a combination of high-producing stock with a high level of flock management and effective disease control.

Disease control and flock management are both intimately associated with pasture management, which in turn determines very largely the quality and cost of the feed supply. This last point is important, because the efficiency of sheep as food converters is such that the fullest use must be made of grass if home-produced lambs are to meet the competition from imports.

The battle of fat lamb production from special forage crops was fought and lost in the closing decades of the last century, when arable sheep farming went out of fashion. The only survival is the use of crops like rape for finishing store lambs, or roots for wintering hoggets and breeding ewes. These later crops, particularly turnips, have again become very important with the advent of precision drills, pre-emergence sprays and mechanical lifters.

GRASS GROWTH AND FLOCK APPETITE

It was pointed out in an earlier chapter that of all grazing stock the fat lamb flock had the best synchronisation of appetite with grass growth. This is a feature which must be exploited to the full if the most is to be made of grass. Fortunately in Britain we have the right material for doing this, in the shape of very fecund fat lamb mothers, either crossbreds like the Mule or the Scotch or Welsh Half-bred, or purebreds like the Clun or the Kerry Hill. A well-managed cross-bred ewe weighing 65 kg will produce her own weight in weaned lambs and 3–3.5 kg of wool annually.

Except for the out-of-season product from a breed like the Dorset Horn, the aim should be to lamb about a fortnight before active pasture growth commences—towards middle of March in the south and west, and towards early April in northern districts. The intention should be to have lambs which are big enough in April and May to make full use of the grass flush.

This will necessitate supplementary feeding, but this can be reduced appreciably by thoughtful planning. Some fields which were eaten bare in the previous October should have been rested for the whole of the winter so that there is appreciable cover on them in March to receive the ewes from the lambing field. An alternative is an area of Italian ryegrass which has been similarly prepared and which has received an early dressing of about 50 kg N per ha.

One disadvantage of Italian is that it will poach badly in a wet spring, and the ewes and lambs will not have a dry lie unless they can run back on to an adjacent old pasture. In most years, however, it is a wonderful standby and provides a breathing space for other pastures to grow some cover before they are grazed in their turn.

DISEASE CONTROL

If it can possibly be managed, the pastures which ewes and lambs are grazing should not have carried this class of stock in the previous year. This is especially true in those districts where Nematodirus is a problem—in areas where grass growth tends to be late in starting. In the absence of a clean-pasture approach to control, it will probably be necessary with intensive stocking to adopt preventative dosing with one of several specific drenches that are available.

Alternatively, if the intensity of sheep stocking is not high, they should be spread thinly over the pastures along with cattle to dilute infection and to control grass. Extremely good lambs can be produced by this method, but the level of output is low and so it is only suitable for large farms with low rents. This is the traditional method of fat lamb production on large farms in Northumberland, where the output seldom exceeds ten fat lambs and a fat beast per ha.

Whatever the system of grazing may be, it is all-important for ewes and lambs that the pasture is short and leafy. Sheep do not thrive on long pasture. The explanation is largely nutritional, but there is also a much greater tendency towards lameness on long pasture—not necessarily attributable to foot rot, but rather to scald. Whatever the reason may be, it is essential to avoid lameness if lambs are to grow and fatten to the limit of their capacity. Anyone with experience of weighing experimental sheep will need no convincing on this point.

Foot rot is an avoidable affliction, because the organism responsible for it does not live in a free state on pasture for more than 2–3 weeks. This fact was first exploited by Australian workers who pioneered a clean-flock routine for eliminating the trouble. The starting point with an infected flock is trimming of hooves and passing the flock at intervals of 3–4 days through a foot-bath containing a solution of formalin. This will clear all but chronic cases, and these should be isolated from the main flock which can then be transferred to a pasture that has been clear of sheep for at least three weeks. The chronic cases are kept in isolation and subjected to intensive curative treatment.

This course of clearance is best undertaken in the summer, when the ewes are dry rather than when they are pregnant or suckling lambs. Apart from the fact that they go into the winter with sound feet, there is a minimum of disturbance. Once the flock is clear it should be kept in this state by routine paring and regular use of the foot-bath. One should try to put all sheep through the foot-bath at monthly intervals, except over the lambing and post-lambing period when disturbance should be avoided as much as possible. The relatively new development of foot rot vaccine does not remove the onus for good flock management. The use of the vaccine may help, but it is not a cure-all and it must be used side by side with regular paring and foot baths.

SYSTEMS OF GRAZING

Turning now to systems of grazing, it is generally recognised that set-stocking will give the best lambs if it is properly worked, but it puts a lower limit on carrying capacity as compared with rotational grazing, with its manifest advantages in planning and conserving surplus feed. Important considerations in making rotational grazing give of its best are the avoidance of disturbances of the flock, especially when the lambs are young, and the maintenance of a good plane of nutrition throughout the suckling period. One just cannot afford to turn fat lambs into scavengers to utilise the last bit of grass in a field. It is better to use followers in the form of dry sheep or store cattle for this purpose or else to use the mowing machine, preferably as part of the conservation programme.

When the flock is being moved from one pasture to an adjacent one, it is better not to drive them but to open the gate and let them filter through. This is especially advantageous if the lambs are young, when there is a great likelihood of separation of lambs from their mothers. One should try and plan that fields used for rotational grazing are in a block.

SALE OF LAMBS

The primary aim should be to grade as many lambs as possible fat off their mothers. Where the flock is lambing in late February-early March marketing can begin as early as the middle of May, because single lambs by a Suffolk or Hampshire ram out of a crossbred ewe can reach a weight of 35–40 kg at 11–12 weeks. It is true that such lambs will reach much heavier weights by the end of July to kill out at 24 kg deadweight, but this involves an inefficient use of food. Lambs at this stage grow at the rate of only 0.2 kg per day, as opposed to 0.45 kg daily in the first twelve weeks of life.

There is still a tendency, especially in the North, to market lambs at very heavy weights. In the long-term there is a very real danger, with increasing ewe flocks, of British farmers becoming their own worst competitors over the July–September period through flooding of the market with lambs at greater weights than the trade requires. Taking the broader, long-term view, it is wise to try and market as many lambs as possible in the May–June period, in order to spread supplies.

There are at least three good reasons for avoiding the marketing of a high proportion of Down-cross lambs some time after weaning. The first is that they generally suffer a considerable

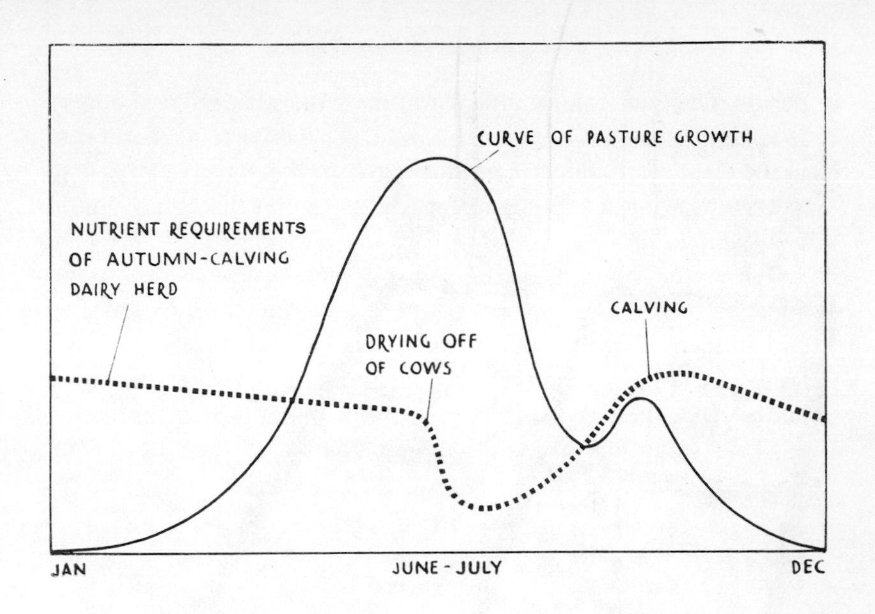

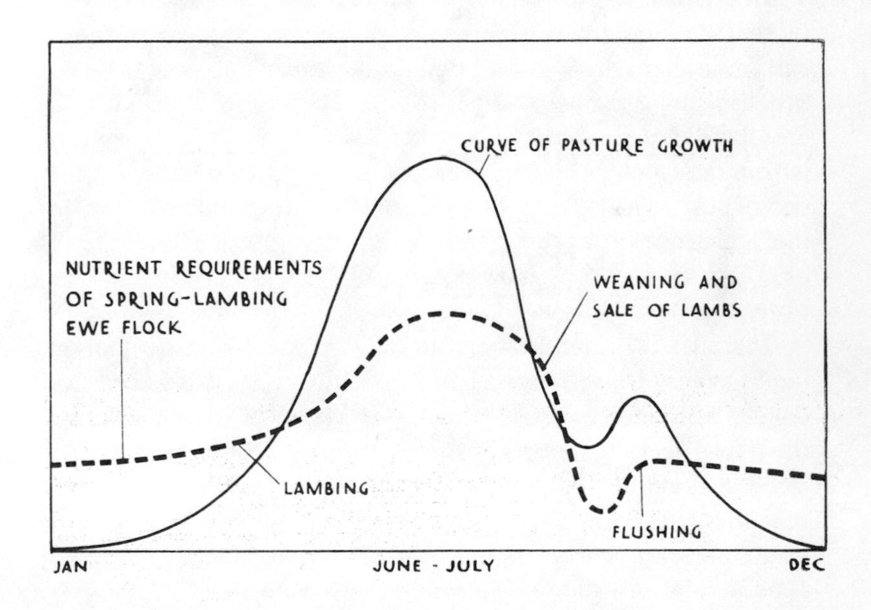

FIGS. 2 and 3

Synchronisation of herd and flock appetites with grass growth.

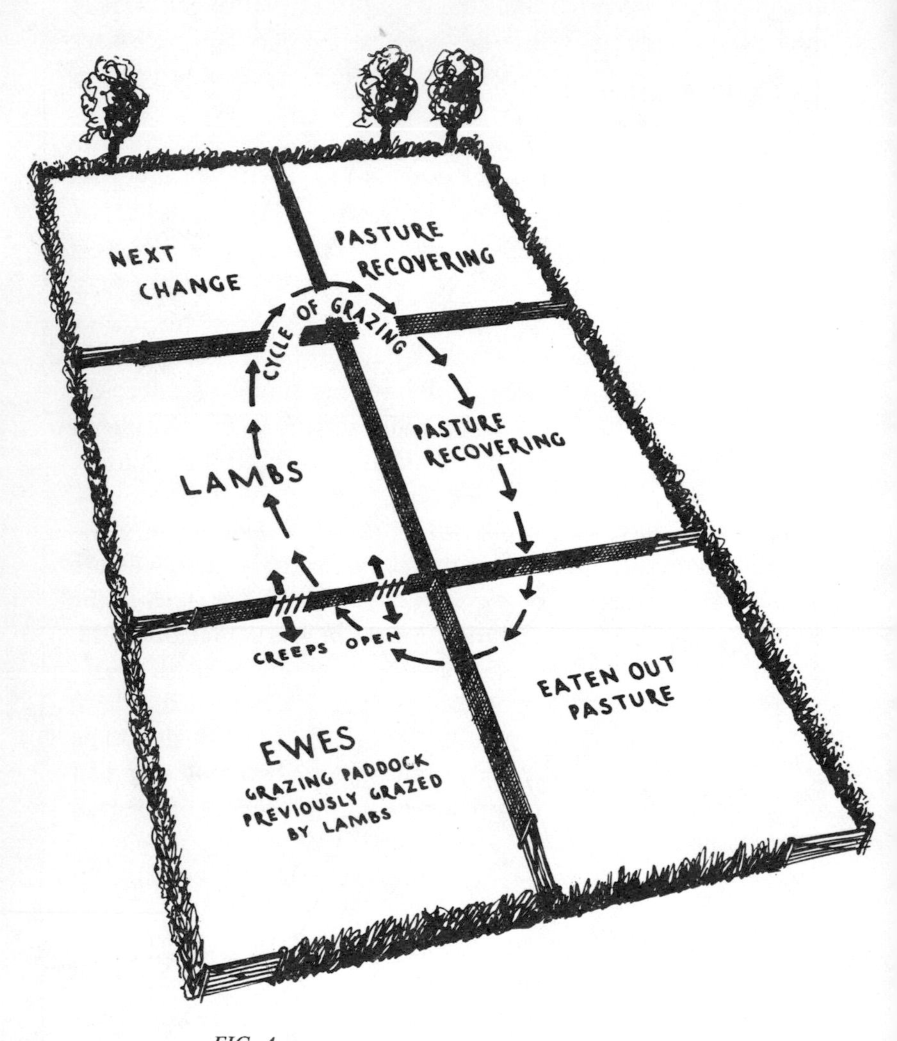

FIG. 4

Arrangements of paddocks for forward-creep grazing.

check after weaning, especially if the summer is wet and the grazing is soft. Worms, both lung and intestinal varieties, play their part in reducing thrift, and generally it becomes necessary to introduce box food, which puts another expense on production costs.

Second, there is a considerable supply of hill lambs, both pure and crossbreds, coming onto the market in the autumn and winter and these are much more attractive to the trade. Those mountains of tallow which are Down-cross hoggets marketed off root crops in the winter represent an extremely wasteful use of food, and they do not make good eating carcases because they are overfat and there is a "woolly" flavour in the meat. From a national viewpoint, if the supply of hill-bred lambs is insufficient for the winter trade, it would be better for the early districts of Britain to expand their production of out-of-season lambs using Dorset Horn or cross-Suffolk ewes.

The third reason is important from the individual farmer's viewpoint, in that every hogget that is carried on autumn pasture and fattened on feeding crops helps to reduce the capacity of the farm to winter ewes. The winter in Britain is a sufficient bottleneck in reducing the scale of fat lamb production, without aggravating it by these extra mouths. It is the fecund breeding ewe which really makes a sheep farmer's income. The aim should be to have every lamb bred on the farm away before the end of September at the latest, in order to build up reserves of foggage for the winter months.

FLOCK MANAGEMENT AFTER WEANING

Even with the best of management with highly-fecund ewes there will be a proportion of lambs that will not be fit for slaughter at weaning. At this after-weaning stage they are much more susceptible to ordinary round worm infections. It is sound policy to drench them with an efficient drug which will clear out most of the infection before moving them on to clean clovery aftermaths where they seem to thrive best on a system of set grazing. Invariably it will pay to provide some trough food in the form of a rolled cereal. There is no need to add a protein supplement for there is ample protein in good grass.

In providing good food for them, one must at all costs avoid contaminating pastures which are intended for the ewes and lambs in the following year. These tail-end lambs can be regarded as suspect on this account at least till the end of October. For this reason, farmers may be well advised to sell their "shott" lambs, preferably to a dairy farmer whose enterprise depends upon cheap store lambs of this kind.

Immediately after weaning, ewes should go on to hard commons to dry off their milk and to prevent any rapid increase in body condition. This is the point where they come into their own as tools in pasture management, cleaning up rough feed, especially that on old leys or permanent pasture. There is another point to this, for one must start stock-piling pasture for the winter before the end of August.

This method of handling ewes pin-points the importance of having good fencing so that they are kept where it is intended that they should be kept. In fact, one cannot contemplate intensive fat lamb production unless fences are really sheep-proof—otherwise the flock becomes master of the farm instead of a working partner.

About four weeks before mating, the ewes must be given a rising plane of nutrition so that they are adequately flushed before they are joined by the tups. It is not good enough to put them on to good pasture the week that mating starts, for there will be insufficient time for good feeding to exert its desirable effects, namely an increase in multiple births and a concentration of lambing into a short period. One can say that a flock has been adequately flushed only if the great majority of ewes have been covered within the first three weeks.

Incidentally, there is little hope of getting a concentrated mating with crossbred ewes with a hill ancestry before the latter end of September. If really early lambs are required, the cross-

Suffolk ewe has much in her favour for she will usually come in season in late August-early September with proper treatment.

Where practicable, individual mating should be adopted with a ram to 50–70 ewes in the one field. If the ram is raddled with colour changes every sixteen days in the order yellow, red, blue, two objects can be attained. The first is a vital check on the fertility of the ram, and the second is that the flock can be divided into early- and late-lambing groups prior to lambing. This means that supplementary feeding can be delayed with the late-lambing group to effect an appreciable economy on this score. It also enables more careful attention to be paid to the smaller groups at due dates.

Once mating has been completed, the ewes can be brought together to work for their living for the first 100 days of pregnancy. Rotational grazing is much to be preferred at this state, because available grass can be rationed into the winter. Their first job can be the cleaning up of stubbles prior to winter ploughing, and then they can go on to pasture, remembering always the proviso that some pasture must be handled in such a way as to provide good feed in the lambing fields, and in those fields to which ewes and lambs will be drafted out of lambing fields.

This is where the stored-up foggage comes into its own, but do not attempt to graze out a foggage field in one period. After perhaps 7–10 days there may be still quite a lot of feed remaining, but it will be fouled and unattractive. A rest of 2–4 weeks will freshen it up and then it will probably carry the flock for another week.

At the Welsh Agricultural College sheep normally based on the beef and sheep farm Tanygraig, move in November to the dairy farm—Frondeg—and remain there until about the end of January. They do a first-class job of cleaning up grass left by the cows as they graze the farm rotationally, spending about a week to ten days in each field. The rest allows Tanygraig Farm to freshen up, and on their return in early February there is ample fresh grass to keep the ewes going until lambing. Two or three new reseed fields are, however, left untouched during this late winter period and become available for the ewes as they lamb. This system could be adopted by "pairs" of stock and dairy farmers to mutual advantage of both. The dairy farmer's surplus grass could be removed in November–January avoiding much winter kill from overgrown clumps of grass and turned into profit by charging the sheep farmer who would be equally pleased to have fresh pastures on the whole of his farm.

SUPPLEMENTARY FEEDING

Every effort must be made to conserve a rationed grazing to put off the feeding of expensive supplements as long as possible. These supplements will include hay, silage, roots and concentrates, according to the circumstances of the farm. If silage is used, it must be of good quality, such as one would feed to high-producing dairy cows, made from leafy material and having a high dry-matter content. Do not force sheep that have never had silage before to eat this food within six weeks of lambing.

It is advisable to start silage feeding well before Christmas with a flock lambing 10–11 weeks later, using a "sacrifice field" for the purpose and providing some good quality hay as well. They will be well settled on to this feed before the critical last third of pregnancy starts, and there is an additional advantage that grazing elsewhere on the farm can be saved for the lambing period. Silage is now popular with sheep farmers who have taken care with both production and feeding.

Sensibly used, silage is valuable in easing the wintering problem, but at the same time it is not suggested that roots should be dropped altogether. Turnips and mangolds, in particular, if fed in moderation just prior to lambing, are a very valuable food in providing succulence. After lambing, these roots are not only useful in stimulating milk yield, but they are a safeguard if the winter is protracted and grass is in short supply. The implication is that one should grow only a small area of roots, but attempt to grow a high-yielding crop.

One should not underestimate the value of the good hay for wintering ewes, and those who are able to grow lucerne will find this excellent, provided the leaf has been saved and it has been cut well in advance of flowering. Good pasture hay with a high proportion of clover is equally acceptable. Poor quality mouldy hay is particularly dangerous for feeding heavily in lamb ewes because usually its feeding is accompanied by 'back body' troubles.

NUTRITION IN LATE PREGNANCY

Even with good hay, silage, roots, and whatever pasture there is, it will be necessary to feed some concentrates in most parts of Britain for 4–6 weeks before lambing and for 2–4 weeks after lambing. This can be crushed oats and/or barley, with some groundnut or soya bean meal in the proportion of 4 or 5 to 1, according to the amount and quality of available grass. Feeding may start at 0.1 kg per day per ewe, rising to a maximum of 0.4–0.5 kg daily at lambing and for a fortnight after, with a tailing off as pasture comes away. This means approximately 20 kg of trough food per ewe, under average conditions.

It is unwise, however, to lay down hard and fast figures as these are dependent on size of ewe, conditions of the flock, weather conditions and time of lambing. The all-important object is to ensure that ewes are improving in condition as they come up to lambing. As Wallace demonstrated with his classic work at Cambridge in the 1940's, one can afford to keep a ewe lean over the first two-thirds of pregnancy, but she must be on a good plane over the last third to ensure strong lambs and a good milk supply. His results were just as good with ewes on this regime as they were with sheep kept on a high plane throughout pregnancy, but the cost of keeping them was obviously much less.

Above all else, Wallace emphasised the danger of getting ewes too fat early in pregnancy and forcing them to live on their condition as lambing drew near. Such treatment increases the danger of twin lamb disease, and results in small weakly lambs and an impaired milk supply. Apart from wool, the lamb crop is the only produce of the flock, and if this is poor the profit will go. One has to reckon these days on the first 120 lambs per 100 ewes covering costs of production, but every lamb reared above this means increasing profits. With the fecund breeding ewes used in Britain, no-one should be satisfied with less than 150 weaned lambs for every 100 ewes put to the ram.

Wallace's work contributes greatly to our knowledge of achieving this target and, above all, to the wisdom of controlled grazing of the ewe flock in the early part of pregnancy along the lines which have been indicated to avoid the danger of feast and famine. Undoubtedly the biggest problem facing us in fat lamb production is this one of wintering a large flock cheaply and well. It is quite feasible to have 8–12 ewes per ha in the spring and summer, and still have elbow room for conservation and the running of cattle. The real headache comes in wintering these sheep without having to adopt expensive cropping or supplementary feeding.

For many years now the Cockle Park and Nafferton flocks of the University of Newcastle have both been in-wintered. The greatest benefit to both farms is not the advantage to the ewe herself, who after all has her own very effective top coat and can be fed silage, hay or roots outside, but the tremendous benefit achieved by keeping the ewe flock off the pastures from the turn of the year to lambing in late March, allowing the grass land to rest and above all preventing poaching, itself a serious drawback to producing spring grass. Care must be taken not to spend vast amounts of money on pre-cast concrete and steel palaces when cheap 'lean-to' type pole barns are equally effective. In most instances, however, a dry sheltered field or waste ground can be equally effective for wintering sheep, its only drawback is that shepherding is less pleasant in wintry weather than in a building.

The arable farmer with stubbles and sugar-beet tops has these to eke out his pastures, but the predominantly grassland sheep farmer must plan his grazing and conservation with the greatest care to safeguard the health of his flock and of his bank balance. Nevertheless, it is a very straightforward kind of farming, and it is surprising that more farmers are not undertaking it as a main line of pasture use. We know, however, that many visiting New Zealand sheep farmers are grateful that this is so, because they recognise the competitive pressure that would be created if we really got down to the job of producing fat lamb from our grassland. Unquestionably we have the advantages of better fat lamb mothers, and the ability to market a fresh product at substantially heavier weights than New Zealanders dare send to Britain.

Possibly, as a New Zealander, the senior author should let sleeping dogs lie and not write about such things, but he feels less of a traitor when he remembers how strongly entrenched is the view that a sheep's worst enemy is another sheep and the conservatism this engenders. It used to be true before advances in veterinary science gave us a better understanding of how to control disease in sheep, but it is not true today for the man who uses this knowledge to full advantage.

CHAPTER XVII

FORWARD CREEP-GRAZING OF LAMBS

THE first experimental work with the forward creep grazing of lambs was undertaken at Cockle Park in 1955. A straight comparison was made between ordinary rotational grazing, where the ewes and lambs run in the same paddock with a change every 3–5 days, and rotational creep-grazing where the lambs but not the ewes have access to the next paddock of the rotation.

After two years of comparison it was abundantly clear that the creep system was much superior. The lambs in the second year of the trial had an advantage at sixteen weeks of 5 kg over their mates on ordinary rotational grazing. This difference was reflected mainly in much better condition, for the great majority of the creep lambs were fit to grade whereas most of the control lambs were forward stores.

This superiority of the creep lambs was attributable to two main factors. First, they had a lower worm burden. Possibly this was because they grazed the top of the pasture while the ewes, with a measure of age tolerance for parasitic worms, ate into the sole of the sward and in the process helped to destroy infective larvae. Secondly, they were continuously on a good plane of nutrition because they always had access to good grass. Under ordinary rotational grazing they tend to be on a saw-tooth plane of nutrition, because there is a progressive fall in the quality of grazing up to the point where a paddock is eaten out and the flock is given its next change.

Lambs did not have this fluctuation with creep grazing, but the ewes were subject to it. As a consequence of the rather poorer quality of grazing they were offered, the ewes weighed 2.5 kg less at weaning than those on the control. There was no disadvantage in this because they were more easily dried off and were prevented from getting over-fat prior to flushing.

COMMERCIAL APPLICATION

Since this original experimental work, creep-grazing has been used commercially on two College farms, using flocks of about 200 ewes plus their lambs. A number of practical problems in applying the technique have been ironed out, and it is now possible to state that when the system is worked properly it is a most effective method of intensifying ewe and lamb stocking without detriment to the quality of lambs. It is not suggested that creep grazing will give better lambs than those produced from set stocking, where cattle are used to control surplus pasture, but that it is capable of producing 30–35 fat or very forward store lambs to the ha on good pastures. For a number of years on our Nafferton farm we have consistently averaged over 900 kg liveweight of lamb per hectare in a suckling period of about sixteen weeks. The lambs have been drafted as they reach 38 kg liveweight, and a very high proportion sold before the normal weaning age of 16–18 weeks.

The recommended pattern for the system is a division of the unit into six paddocks which may be grazed cyclically. A rectangular field of, say 12 ha is divided down the long axis, and there are two additional cross fences to give the six enclosures of 2 ha each, according to the plan given in Figure 4. The fencing need consist of no more than sheep netting and stakes with light strainers at the corners, provided cattle are not grazed as well. If they are, the netting fences will have to be protected by an electrified wire.

Creep gates are inserted at 50 m intervals—the normal length of a roll of sheep netting. Possibly the best type of creep gate is the Stanton creep hurdle with an adjustable second-from-bottom horizontal bar. If the uprights of the gate are slotted and butterfly nuts with bolts are used, this bar can be quickly moved up and down to permit or prevent movement of lambs. Experience will soon determine the amount of gap necessary in order to allow the lambs, but not the ewes, to pass through the creep. With strong crossbred ewes, a 40 cm gap is the maximum that can be allowed, and even with this the odd ewe may pass through. If she does, this does not mean that the system has broken down.

Normally, ewes are kept in a paddock for 3–4 days and then they are moved to the next paddock to which the lambs, meanwhile, have had access. The lambs, in their turn, pass through the creeps into the next paddock, which will have fresh grazing ready for them. The whole rotation takes 18–24 days, and for two-thirds of this time any one paddock is rested to allow a recovery of growth. The aim should be to try and adjust stocking so that each paddock is completely eaten out in its turn and recovers to the point where there is clean fresh grazing when the lambs are given access to it.

Assuming a start has been made at the end of March, feed will tend to get ahead of the stock about the beginning of May. It may then be necessary to by-pass two of the end paddocks from the grazing cycle and to cut these for a light crop of silage. If this is done, the crop should not be allowed to get mature and develop a yellow bottom, because recovery will suffer. Cut early to ensure a green bottom to the sward and the right continuity of grazing.

MANAGEMENT FACTORS

Advisedly, a field which is used for creep-grazing should not have been used for fat lambs in the previous season. This is part of the clean-field technique which is advocated as the most satisfactory method of controlling Nematodirus.

The field should be grazed bare in the autumn to remove any rough growth, but in sufficient time to allow a fair amount of recovery before the winter sets in so that there is some grazing available for the ewes when they are drafted on to the area in the early spring. This is a most important precaution if one is lambing a month or more in advance of normal pasture growth, so that both ewes and lambs have some pasture to augment any hand feeding. Usually it is advisable to apply some nitrogenous fertiliser say 70 kg N per ha, as an early-bite dressing some 2–3 weeks before grazing commences.

It is most important not to delay the start of the system in the spring, for the lambs should learn to use the creeps when they are 2–3 weeks old. For a start they will just go through to play, but they quickly learn to graze the fresh growth. Many failures with creep-grazing can be attributed to a delayed start to the system, because a lamb 4–6 weeks old is accustomed to grazing beside its mother and it seldom learns to creep satisfactorily. It is wise to erect the fencing in the winter, so that the unit is ready for the ewes and lambs as they are drafted from the lambing field.

Some farmers, particularly those who are lambing early in order to catch the high-price market for spring lamb, offer trough food to the lambs in the creep paddock. This is not really necessary to teach the lambs to creep, and is justified only on the grounds of improving their plane of nutrition in order to catch an early market.

In the early stages—during the first six weeks—the criterion in moving sheep will be the needs of the ewes, remembering the importance of a good initial milk supply in getting thrifty lambs. Each paddock should be eaten out well in its turn, but the ewes should not be punished to achieve this end. Progressively after the six weeks' stage the ewes can be made to work harder for their living, so that they become scavengers behind their lambs as the pasture tends to run to seed in late May and June. There should be no need for topping if stocking is adjusted properly. Nevertheless it must be resorted to if there has been a failure to keep on top of grass, for it is essential that the lambs have fresh short growth each time they move to a new paddock.

We found that we had very little trouble in drying off the ewe's milk or getting the lambs settled after weaning. Properly worked, the last stages of the system can be regarded as part of the weaning process.

One of the principal reasons for creep-grazing not being successful is a failure to stock with a sufficient intensity so that there is enough pressure on lambs to encourage them to creep. If one is running only ten ewes and 15 lambs to the ha, the lambs will probably find sufficient attractive food in their mother's paddock and they will not use the creeps freely. At double the stocking rate there will be this pressure, and a good pasture will stand such a rate of stocking quite comfortably. In fact, one of the features of the system is the tremendous recovery that takes place in the 14–16 day rest period, and unquestionably this is due to the stimulus the pasture receives from the heavy return of excrements.

Normally it does not pay to keep the system going after the beginning of July, assuming a lambing in early March, for by this time the bulk of the lambs will have already been sold fat off their mothers. There is also the practical difficulty that the remaining lambs will be almost as big as their mothers, which will be shorn by this time, and so there will be increasing difficulty in keeping the ewes from using the creeps.

POTENTIAL OF CREEP-GRAZING

Forward creep-grazing has not caught on as was originally thought, but no apology is made for its inclusion here. In 1961 the senior author stated that whether it ever was to get past the juvenile state depended on the relative profitability of fat lamb production and the preparedness of farmers to take the necessary steps to make the system work. There is no doubt in our minds that sheep production is entering a very profitable era, and that the intensification of production that is possible with creep grazing will make it an economic proposition particularly in the Common Market with its demand for leaner and lighter carcasses than those we have been accustomed to producing. A number of farmers have attempted creep-grazing and discarded it because "it has not worked". The system does not work on its own accord, for it has to be properly managed. Invariably there are obvious reasons for failure—for example, starting the system too late in the season and, above all, understocking.

There must be sufficient grazing pressure in the ewes' paddock to induce lambs to seek the fresh grazing through the creep. At the same time, one must not press the ewes early in lactation to the point where milk yield suffers, nor is there need to when grass is still leafy and palatable. It is reiterated that there is no point in attempting creep-grazing on reasonable pasture with a stocking rate of 10–12 ewes plus lambs per ha. Here one is better off with set stocking. Another reason for lack of success is a failure to provide the ewes with adequate drinking water. Their needs for water are very considerable during the suckling period, especially if conditions are dry.

There are two sets of conditions where creep-grazing is likely to be of special value. The first relates to farmers with a Nematodirus problem who wish to control it by running their ewes and lambs on pasture that did not carry sheep in the previous summer. The concentration of stock which is made possible by creep-grazing means that a large area of grassland can be rested over the critical period from April to the end of October so that it will provide clean grazing for lambs in the following year.

The second possibility for creep-grazing is on the medium-sized partly-arable farm where it is intended to produce fat lambs rather than milk. It would be possible for a farmer with 50 ha of grass to carry up to 500 breeding ewes plus their lambs on half of this grass during the spring and summer, with the balance of pasture being used for conservation and for such cattle as the farm might carry.

The whole of the grassland would be available for the wintering of the ewes together with stubble and other feed derived from the arable section of the holding. It would involve both the systematic saving of foggage for the back-end of the year, which would be used by rotational grazing, and the supplementation of ewes in the winter with silage and hay as well as some concentrates over the lambing period.

With good management and reasonable price levels for wool and lamb such a system, combined with a sensible integration of cattle, has a considerable measure of attraction to someone who does not want the daily grind of dairying. It is true that gross returns per ha will be much less but labour, capital, and supplementary food inputs will also be much lower. Certainly it is a safer form of farming than continuous barley growing on land that by its nature should be worked on a system of alternate husbandry.

One further impressive point about creep-grazing is the tremendous upsurge of growth one gets on the creep-grazing area after the flock has been removed from it. This is primarily due to the fertility which is mobilised by the heavy stocking. In most years we find with the removal of stock early in July that a heavy crop of silage can be taken in August, and this is invariably followed by a strong autumn growth which is useful feed for the ewes at mating time.

We maintain these creep paddocks until after mating is completed, for they are such convenient enclosures for this purpose and provide a means of ensuring individual ram use with a consequent check on fertility.

CHAPTER XVIII

BEEF FROM GRASS

THE explosive increase in cattle prices, either as calves, stores, or finished beasts has caused a major interest in beef production. This interest has been increased by the demand for grazing animals in many arable areas which can only be said to be suffering from exploitation or "mining" as a result of continuous white straw cropping. The price of cereals and the price of calves at the present moment has virtually sounded the death knell of barley beef production. Fortunately this may be regarded as a development that has released more beef animals to be reared and finished at heavier weights on a system based on grass or grass products.

With the enormous increase in the price of land and the increase in price of breeding stock, the simple fact that the production of single-suckled calves is not sufficiently intensive to provide a good living for any but large-scale farmers on relatively cheap land. Even here the production of stores is integrated with other forms of farming, such as sheep on mainly grassland farms, or cash crop production on tillage farms where there are arable by-products as well as some grassland. Generally, however, on this latter class of farm the preference is for feeding rather than breeding stores, and this is right because store stock should come from cheaper rather than dearer land.

The foregoing does not deny that there are a large number of beef-type cows in Britain which are kept solely for producing stores. A large number of the herds used for this purpose have been established or substantially expanded during the past ten years on marginal land which has received considerable Government support in the form of land improvement grants and subsidies. Today, many farmers who have expanded their beef production in this way, and who have created an efficient organisation, find this a worthwhile enterprise which is complementary to, rather than competitive with, their sheep production. However, the fact that it has been necessary to provide such substantial subsidies indicates the economic frailties of this type of farming. This fact must be very much in our minds following our entry into the European Economic Community and the reorganisation of our production grants.

PRODUCTION PROBLEMS

In our context, we are concerned with the problem of feeding these cattle as cheaply as possible and, at the same time, turning out a class of weaner which will command a good price at the autumn sales. Although basically this is a question of using grass and grass products to best advantage, there are other important management points, such as using breeding cows which are adapted to the conditions of the farm, and the timing of calving so that the aim of producing a well-grown weaner is not achieved at the expense of too high a feed bill.

So far as adaptation is concerned, the simple maxim is—the harder the conditions, the hardier should breeding cows be. This means on poorer upland grazings, breeds and crosses like the Galloway, Welsh Black, Blue Grey and Cross Highland. On easier country the emphasis moves to a big-framed, deep-milking cow of the type once produced in considerable numbers by Eire—generally a crossbred by either a Hereford or an Aberdeen Angus bull out of a dual-purpose Shorthorn cow, although this type of animal is becoming increasingly more difficult to obtain. The Hereford × Friesian cow is probably the best available substitute.

Time of calving also varies with class of country. Someone who is able to run the bigger type of cow can usually afford to calve during January and February to produce a well-grown weaner for the October sales which will command a premium from buyers who finish this class of animal as baby beef at 13–17 months of age. The farmer with the hardy type of cow will more usually arrange calving for the February–March period, but there will be an inevitable proportion of later calvings.

There has been a move in recent years towards autumn calving, again with a view of turning out a large-framed store which will be acceptable to the yard feeder some twelve months later. This, however, means fairly expensive winter feeding in that the calf is at least six months old before it goes on to grazing, and there is also the extra feed necessary to keep cows in milk over a six-month winter period. When one considers the difficulty of getting cows in calf in the middle of the winter as compared with the spring, and the consequent spread of calving dates which results in a much more uneven line of weaners, the economics of autumn calving becomes very doubtful. The more one can keep herd appetite in line with the availability of grazing, the better will be the prospect of getting a low-cost structure into the production of suckled calves.

WINTER CALVING

Calving in late winter some six weeks before the anticipated onset of growth of grass gives a very satisfactory solution to many problems. In the first instance it decreases the winter fodder and housing needs as compared with autumn calving. Cows calve on winter rations and the flush of milk coincident with the spring flush of grass does not begin until the calf is old enough to take this quantity of milk.

On the whole, back-end grazing seems to be of greater value to single suckling than early-bite grazing. Certainly this view is fairly widely held by Northumbrian and Scottish Border farmers, who are probably among the most efficient there are in this branch of farming.

This arises because of several considerations. The first is that in late districts—and many of the main suckling areas have a late spring flush of grass—early-bite is uncertain and is expensively produced. Autumn growth, on the other hand, is fairly well sustained and generally it is possible with the help of a late-summer application of nitrogen to save foggage for early winter use and so ease the burden of supplementary feeding.

A second consideration is that rapidly growing grass in the early spring is very suspect on this class of farm with regard to hypomagnesaemia. Normally the nutrition of single-suckled cows in the late winter is kept at a fairly low plane, and it may well be that this puts them on something in the nature of a knife edge, so that any unfavourable stimulus may precipitate the disease. This need not necessarily come from young grass, for there is good evidence to show that the onset of tetany in an out-wintered suckling herd may be associated with the occurrence of very cold spells of weather.

A third factor contributing to the present emphasis on late autumn rather than early-spring grass is the increased popularity of silage for the feeding of suckling cows over the period of insufficient grazing. This has followed the adoption of self-feeding or easy-feeding of silage, especially where cows are yarded.

Where cows are outwintered, however, most farmers prefer not to self-feed, for the good practical reason that there is invariably a morass at the silo. For this reason, and because most farmers prefer to limit silage feeding to about 30 kg per day in order to encourage cows to eat out any rough grazing that may be left, hand feeding of silage is more common with out-wintered herds.

Where there is a considerable snow risk, it is prudent to have a reserve of hay in case silage feeding is impossible. In fact, on the high hills, hay is a much more practicable proposition than silage as a main winter supplement because of difficulties in handling silage in deep snow.

GRAZING MANAGEMENT FOR SUCKLER COWS

Feeding during the summer months presents few problems apart from those of husbanding grass to avoid waste, without upsetting the milk supply of the cows or limiting in any way the calves' very considerable capacity for growth. Grazing management will vary considerably according to the nature of the farm. On poorer hill pasture the herd should preferably be allowed to range freely, generally in association with sheep, but on better land where mowing is possible some form of long-interval, rotational grazing combined with conservation will give better results.

Though most farmers practise mixed stocking through the grazing season to achieve a dilution of worm contamination of pastures, many prefer to keep sheep and cattle separate as far as this is reasonably possible. Sheep tend to foul grass for the cattle, and certainly they compete for the short leafy grass which calves prefer. This does not demand complete separation of the two species over the whole of the grazing season, except where the object is to develop a clean-field programme for Nematodirus control, but rather the practice of giving a short rest period after sheep have been on a field before the cows and calves are turned in. After the calves have been weaned the situation is different. As pointed out previously, this is a time when cows can be used to clear rough grass.

Where there is a programme involving a systematic laying up of foggage with the intention of using it as long as possible into the winter, some form of restricted grazing is advisable. Usually fields are so large on this class of farm that this will involve the use of the electric fence, though not necessarily a daily shifting of the fold fence. Because this class of animal is not used to close restriction, it is possibly better to give them larger breaks which will last them 3-4 days.

This leads on to the question of the best sort of pasture for the production of suckler calves where these are a main line of production rather than a sideline, as they may be on a cash cropping farm. On economic grounds, one can say very firmly that pastures should be permanent, or at the least very long leys, for the cheaper the grass the greater will be the hope of a reasonable profit. By the same token, the manuring should be such that the greatest use is made of cheap clover nitrogen. In other words, the endeavour should be to make good any lime, phosphate and potash deficiencies where these occur, and to use nitrogen very sparingly for stimulating crops of grass that are intended for conservation, either as hay or silage or *in situ* in the form of foggage.

There is another important practical consideration in this preference for old grass, namely its capacity to stand poaching, which is a matter of some concern where cattle are outwintered. A farmer is fortunate if he has some rough, well-sheltered and fairly dry land that he does not mind punishing, for this will be almost as good as housing to him and considerably cheaper. Housing of the whole herd will be necessary under some conditions, but on most single-suckling farms only limited housing will be available. This will usually be reserved for older cows and especially those in poorer condition which require preferential treatment.

Where pastures have to be sown, there is much to be said for a Cockle Park mixture, with its durability and its dual-purpose function, in the sense that it gives a reasonable quality of grazing and a considerable amount of bulk. Its cocksfoot component makes it very suitable for putting up for foggage, provided some nitrogen is applied in the late summer, not only to provide greater bulk but to reduce the amount of winter burn. One point to remember in using a pasture containing cocksfoot for foggage is that this species can easily become dominant to the exclusion of ryegrass and clover if it is repeatedly closed for this purpose. It is wise to ring the changes to preserve a balanced sward.

Latest MLC figures show that major factors that determine profitability are stocking rate and liveweight gain of the calf. The latter has been improved in recent years by the greater use of dairy cross cows that are mated to heavy breed bulls like the South Devon, Charolais and Simmental.

FEEDING SUCKLERS IN THEIR SECOND SUMMER

We are dealing here with suckler calves mostly born in the spring of the previous year and about 13–14 months at turnout in April. These will have been on a winter diet of silage and hay and 1.5–2 kg of cereals per day, and will have gained some 0.5–0.6 kg daily during the winter months to weigh at turnout some 300–360 kg for bullocks and 280–310 kg for heifers. The aim is to provide good grass for these animals to achieve a daily liveweight gain of a kg, the bullocks finishing after six months at about 500 kg and the heifers rather earlier at about 420 kg.

If possible, the transition from yard feeding to grass should be fairly gradual. It is unwise to push these yearling stores on to grass too early in the season before there is ample grass which has hardened a little. If they have been in a completely covered yard, which is not necessarily the best winter accommodation for store stock because it makes them a little too soft, they should first go on to a field with good natural shelter until they are acclimatised.

Decisions regarding disposal at the end of the summer have to be made very carefully because the autumn is the lowest-priced period, although at this time of the year quality beef of the sort that is being described usually earns a considerable premium over coarser types. Should the conclusion be reached that it is preferable not to sell 500 kg bullocks in the autumn, even if they are fit, but to retain them for the rising Christmas market, it is important not to leave them out on grass to the point where they start to lose flesh. This is more likely to happen in a wet autumn when grass is very soft and low in energy value. Such stock should be brought into the yards just before the critical stage is reached; here they will continue to grow and put on condition on a diet of silage, hay and 2–3 kg of crushed cereals, provided the hay and silage are of a quality appropriate for fattening.

Today it should be the aim with all specially-bred beef stores to slaughter them in their second year at any stage from 14 months onwards, either out of yards or off grass, but there will always be a proportion which are not finished by two years. This is most likely to be the case with the smaller weaners of about 210 kg bred from hardy hill cows. The choice with these is either to sell them fat as a second crop out of the yards at the end of their second winter at ages of 24–26 months, or to winter them more cheaply and finish them off pasture in the May/June period when beef prices are usually high. The emphasis with this class of stock, which have to live through two winters, must be on very cheap feeding.

Incidentally, these 2-year-old stores appear to be more thrifty on very good old grass than younger animals. It is for this reason that many farmers with fattening pastures such as those of the Midlands often prefer Irish stores, which usually have some age on them.

It has been pointed out in an earlier chapter that there seems to be no point in grass fattening, of adopting intensive grazing methods. Cattle feed better on a system of set stocking, provided herbage is kept in a leafy condition. Pasture control will have to be effected mainly by adjusting cattle numbers or sheep numbers and by the use of the mowing machine if there is a tendency for swards to run to head. As grass begins to deteriorate, it is generally possible to reduce stocking pressure, either by sales of fat cattle, or by bringing in aftermath fields.

The art of good stockmanship in this style of farming is to ensure that feeding animals are able to fill their bellies quickly and to lie down and cud, rather than be on the move because they are not satisfied. They should, therefore, be subjected to the minimum of disturbance, consistent with good pasture management.

BEEF FROM THE DAIRY HERD

Some 75% of our beef supplies owes its origin to the national dairy herd. These calves are based mainly on the Friesian, with a very large proportion being pure bred. The importation of larger bulls, the use of our own larger breeds, and the enormous increase in calf prices has, however, resulted in calves for beef being bred from the smaller breed of dairy cow. Bulls such as Charolais, South Devon and Simmental have been used successfully to produce a very reasonable calf from the Ayrshire, the Guernsey and the Jersey cow. The Cockle Park Jersey herd has produced some very good beasts by a Charolais bull which have reached weights of over 500 kg at 18 months of age on a semi-intensive system.

THE 18-MONTH OR SEMI-INTENSIVE BEEF SYSTEM

This system of beef production has grown to be a national system. In the first instance we will consider the autumn-born calf, that is born between the months of July and November. This is appropriate as this is the time when availability of calves from the dairy herd is most plentiful.

Many arguments will always take place as to when to buy calves. Rearing is easier in the warmer earlier part of the season unless of course first-class controlled-environment calf houses are available. On the debit side of course the earlier bought calf has a longer and more expensive first winter, although he normally reaches a higher slaughter weight and a higher gross return. This is an important factor when the initial calf price usually forms a very high proportion of the eventual selling price.

The following programme is one that can be recommended as a well tried system with reasonably good targets:

	Stage	*Weight (kg)*	*Diet*	*D.L.W.G./day (kg)*
Oct-Nov	Birth to 5 weeks	40–65	Colostrum for 5 days. Milk substitute for 30 days. Hay and early weaning mixture to appetite	0.6
Nov-Dec	5 to 12 weeks	65–110	Hay—early weaning mixture to appetite	0.7
Jan-April	12 to 28 weeks	110–225	Hay and/or silage to appetite. Concentrates at 2–2.5 kg daily	0.8
April-Sept	28 to 52 weeks	225–340	Good clean grazing 1.5–2 kg cereals daily for 2 weeks in spring until 3 weeks before housing	0.85
Oct-March/ April	52 to 78 weeks (slaughter)	340–550	Good silage to appetite. Average feed 3 kg cereals daily (2 kg in autumn, 4 kg in spring)	0.85

Up to the present time this system has had many advantages. Calves were cheapest in autumn, but this is no longer the case. Spring prices for beef have traditionally been at maximum when these animals were sold in March/April and into May.

A well-known master and pioneer of the art of 18-month-beef production is Fenwick Jackson of Shoreswood, Norham, Berwick on Tweed, Northumberland, probably the county with the greatest expertise in this type of beef production.

Shoreswood is a farm of 325 ha in the Tweed valley with 220 ha of cereals, 84 ha of short leys and 20 ha of unploughable permanent grassland. On this area of grass some 300 finished bullocks are produced off the farm annually, of which 250 are autumn born and marketed at 18–20 months of age. This enterprise is designed to play an important and profitable part in the total farm business, using to the full the grass break of perennial and Italian ryegrass.

Good quality Friesian calves are always bought and these are early weaned and fed according to the system described earlier, for the first winter.

In the spring of 1972, 242 were sold at an average weight of

555 kg at 600 days, having achieved a liveweight gain of 0.86 kg throughout their lives. Approximately one-third of the grass area is grazed until the first-cut silage is removed. The area that has been spring grazed and part of the first-cut acreage is then shut up for second-cut silage. This releases a fresh area of grazing for the growing animals and also gets over the fouling problem often associated with this system. After second-cut silage all the grass area is normally available for grazing unless some is needed for a third cut. Mixtures of RvP Italian ryegrass and S 24 perennial ryegrass, or more recently *Barlenna*, are used for these two-year leys and are normally established by undersowing under a barley crop. Although every effort is made to provide clean grazing every year, inevitably cattle, especially on the permanent grassland, use land that has been used in previous years for beef production. To protect these cattle, especially the late-autumn-born calves from husk, an oral vaccine is administered before the calves go out to grass. Cattle are also dosed with a suitable anthelmintic in mid-July to control stomach and intestinal worms.

Considerable applications of nitrogen are used together with phosphate, a large proportion of which is applied as basic slag and potash. The average usage of N, P and K for the period 1968–71 are given below:

	1968–71 average
N	286 units/ha
P	120 units/ha
K	100 units/ha

An important feature of the farm is the quality of the silage offered to the fattening beasts during the winter. Considerable emphasis is placed on obtaining a high dry-matter product that is cut at an early stage and then wilted. The grass is cut with a flail mower as this is known to give a rapid rate of wilting and is then picked up with a precision chop forage harvester. Silage is made as quickly as possible and is immediately sealed with a polythene sheet.

The result of this highly digestible palatable silage is high intakes and very good liveweight gain. In the 1971/72 winter 244 animals averaged a daily gain of 1.0 kg, which is well above average.

Although the system described above fits well with the autumn-born dairy bred calf, we must not forget that a very large number of calves are born in the spring. Indeed a move to spring calving, especially in the West, may produce more of these animals.

At the Welsh Agricultural College's Frondeg Farm spring-born calves are turned out in suitable weather in late spring or early summer, ideally to a silage aftermath. Thirty of these calves born in January/February and early March of 1972 were turned out in May and received 2 kg of rolled barley for 2 weeks and thereafter grass alone, and they averaged a daily liveweight gain of 0.7 kg until housed in November. Fenwick Jackson, also with spring-born calves, feeds 1.5–2 kg barley/day throughout the grazing season, and in 1971 his calves had a liveweight gain of 0.9 kg daily.

This type of animal is yarded in October at 300 kg and is usually fed silage ad lib and 2 kg of mineralised barley, the aim being to achieve a gain of 0.75 kg daily. A strong store of 380–420 kg is turned out to grass in spring. Finishing will depend on breed, and indeed sometimes on type within the breed. Beef/dairy crosses may finish in the late summer, while pure Friesians will often keep growing without finishing and they have to be housed for sale in the November/December period at weights of 500–550kg. At Shoreswood 65 of these animals were sold at 540kg having grown at 0.9kg per day during their final summer with an average lifetime average liveweight gain of 0.8 kg. This type of beef production usually means retaining the animal to a slightly older age of 20–22 months as compared with the autumn-born 18–20 months.

CHAPTER XIX

MILK FROM GRASS

FOR many years we have had in this country two dramatically opposed points of view regarding milk production. On the one hand there is the producer who gets all his production from the cake bag and mainly uses his pastures as exercise grounds. At the other extreme is the grassland fanatic who firmly believes that grass in September will support a cow producing 22 litres per day, retaining sufficient condition on her back to enable him to get her in calf again.

These are extreme examples, both of which have been caught on the band-wagon current at one time or another and have never been able, or have had the desire, to get off it. However, the point we would like to make is that the middle of the road is often the wisest course. The sensible use of concentrates with good grassland management and an intelligent use of fertilisers is the way to reasonable profit from the dairy herd.

In the 1967 edition of this book the senior author stated there may be a fall in individual production as stocking intensities increase, and this is acceptable if production per hectare rises.

This is generally the case, but now, with the cow values almost double what they were in 1967, we must stress the importance attached to production per cow as well as production per ha. A Friesian cow must achieve an output of at least 4,500 litres per year to give a sufficient margin to pay ever increasing costs and leave a margin of profit.

CONCENTRATES—HOW MUCH?

The most important question that an open-minded milk producer must answer is, "How much concentrates can I use to give me the most profitable system?" Does he aim for 5,500 litres average with 2,000 kg concentrate usage or 4,000 litres with 900 kg concentrates? At present milk and concentrate prices, and this has been true for a long time, the former is better off financially than the latter, because he is cashing in on his superior stockmanship and herd management ability. Unfortunately, there are too many farmers who record the higher level of concentrate use and the lower level of milk production, largely because both their herd and their pasture management are deficient.

In many respects milk production per ha is a much more satisfactory economic criterion than production per cow. It will be remembered from Chapter VII that New Zealand evidence, obtained by direct experiment as well as from farm surveys, proves conclusively that rate of stocking is a more important factor in determining level of yield per ha than is production per cow, up to a certain critical point where reduced yields per cow through over-stocking are not compensated by the greater stock density.

There is some evidence to show that density of stocking is also an important factor in determining both level of production and level of profit per ha under British conditions. This comes about through two main factors. The first is the operation of the law of diminishing returns at high levels of food inputs; in other words, there is a curvilinear and not a straight-line relationship between levels of nutrient intake and levels of milk production. The second is a psychological reaction, in that the greater the concentration of stock on a farm the greater will be a progressive farmer's effort to grow more feed and to utilise it effectively. Provided these things are done sensibly, further economies will come through a better spread of overheads and a higher output per labour unit.

GRAZING METHODS

Accepting that a high density of stocking is an important factor in determining profit per ha, we turn now to the problem of ensuring that this is achieved within safe limits. The main factors here are the level of production of grass nutrients, their quality, and the system of grazing. The point was established in Chapter VII that some form of controlled grazing, whether it be by the use of the fold electric fence or by close semi-permanent subdivision to permit paddock-grazing, introduces a critical element of budgeting into the planning of food supplies. It gives opportunities for discreet management of portions of the grazing area, and it reduces the danger of feast and famine, a critically important consideration under our conditions because a farmer can build up food reserves in the form of hay and silage to the point where he has confidence in increasing stocking rate.

If he wishes to increase output even more he can then move from a reliance on clover nitrogen to fertiliser nitrogen, provided he has the additional stock to make good use of this additional herbage. Thus, unlike his New Zealand counterpart, who is denied the use of concentrates on economic grounds, he has another tool which, used sensibly, can further assist his efforts to achieve economies of scale.

Turning to the pros and cons of fold or paddock-grazing, the deciding factor in choosing one method or the other appears to be the relative importance of tillage in the farming system and the expense of moving a paddock system from one field to another. It has already been pointed out that the early conclusions of British research workers on this topic were confused by stocking rates. At similar rates of stocking there seems to be no appreciable advantage in production per unit area, either of herbage or of milk, from the two systems.

The paddock grazing system described in Chapter VII has caught on in very many dairy farms in the United Kingdom. Normally each paddock is used for one day's grazing only, and there has been much debate as to the number of paddocks necessary to make the systems work effectively.

WELSH COLLEGE FARM GRAZING

Many points for discussion will emerge from this description of the methods adopted by the junior author on Frondeg Farm, Welsh Agricultural College. It is an all-grass farm of 54 ha and carries 105 Friesian cows, all of which are spring calvers. In addition, some 30 heifers per year are reared with 20–25 entering the herd at two years of age.

At turning out time the cows first of all graze a fairly bare field, so that the minimum of grass is eaten by them in this vital change-over period. This is a field, without subdivision, destined for silage, as the cows would be difficult to contain in a tight paddock at this time. After 2–3 days, they strip graze an early Italian ryegrass ley, sown the previous autumn and then they move paddocks consisting of longer term leys. Thirty-three one day paddocks, each 1 ha, are available for the cows. However, we have not had to use these 33 paddocks in any spring as yet. Some five or six paddocks have been taken for silage in mid-late May but the important point is that they are available should the grass run out. It is always important that any system is flexible, and to remember that the grass is there for the cows, rather than the cows being there to make the grass look attractive to the casual visitor. In a good system, however, these two criteria are met at the same time.

Rather in conflict with the views of the senior author we virtually use a two sward system. One area is set aside for grazing and one for cutting, but both these areas are not sacrosanct from alternative usage for cutting and grazing. There is no harm, and it is not an admission of failure, if the herd has to be moved to an area planned for cutting for a few days in a dry or wet time.

Four cuts of silage, each taken at 4–6 week intervals are taken, starting in mid-May. About 25–30 hectares are cut for first cut, and a similar area for second cut. Due to the decreased grass growth a smaller area is cut for third cut, and only some 15–20 hectares for fourth cut, as by now grass growth is slower, young stock needs are greater, and a greater area is needed for grazing. This fits in well with maintaining a reserve of grass well into the autumn when cows are easing up in milk production. Sheep rotate around the farm from the time the cattle are housed until late January to clear up surplus grass.

This system has advantages and disadvantages. Its main advantage is as a tool in farm planning, particularly useful for an adviser who can explain and get the farmer to implement a fairly straightforward system, giving a fair amount of confidence often needed when adopting a system of higher stocking rates. To a large extent the conservation and grazing area can be planned. The farmer knows that the dairy cows will be on Paddock 12 on May 15th and again on June 4th and June 25th if he has 21 paddocks. It creates a system of strict timing of fertiliser application, which is essential if the system is accompanied by a heavy stocking rate of a cow equivalent per 0.4 ha. A major advantage of this system to us is in saving of labour. The herdsman has only to close the electric gates after 24 hours and open up the next paddock, a job that literally takes seconds, rather than daily moving of electric fences. This job often has to be done by a second man, which can take many hours per week.

DISADVANTAGES OF PADDOCK-GRAZING

There are also disadvantages. It means that a large number of dung pats, and urine patches appear in the whole area with consequent refusal of herbage. Many people, including the authors, have found, however, that with time, due to the increased activity of the soil bacteria and earthworms, these effects become minimal after 2–3 years and dung pats disappear very quickly, cows graze the sward evenly and topping is seldom necessary. A system, however, that takes care of this is that of Mr Stanley Morrey in Wiltshire. He has a large number of paddocks and alternates cutting and grazing throughout the season, thereby removing the fouling affect. Incidentally by this method he is able to harvest herbage that would have been refused by the grazing animal as silage.

Another disadvantage of a fixed number of paddocks, often 21 in number, is that no account is taken of the seasonal variation in growth of grass. This variation can be taken care of by varying the forward distance travelled by the electric fence when fold grazing. In general, however, this paddock system works well, it has enabled farmers, hesitant to intensify, to do so with remarkably good results. One step ahead of this system, however, for the real expert, is "old fashioned" daily movement of the fence with a back fence as practised by Messrs Bushby and Platt described in Chapter XX.

Paddock-grazing has other disadvantages. Setting up the system is not cheap, water has to be provided as cows will not walk far for water, but water is always cheaper than milk. One must be careful also in fertilising and in the management of the system. Generally low phosphate or potash applications per year are required on paddocks intensively grazed, in contrast to very high potash and phosphate requirements for swards continuously cut for conservation.

We do not want to create the impression that all milk producers use paddocks or strip-grazing methods. Herds in arable areas with short leys cannot justify the expense of setting up a paddock system for one year and it is under these circumstances that an intelligent application of fold grazing can pay dividends. Many herds, however, graze a field for 4–5 days before moving on to the next. This has a serious disadvantage in that the cows have a progressive deterioration in the quality of their grazing with a consequent fluctuation in milk yields.

There is a considerable danger when a large field is fold-grazed that the last breaks may suffer deterioration with an opening up of the sward and loss of clover, caused by grazing at a very advanced stage of growth and consequent overshading effects. This will often occur when fold-grazing of a field is repeated several times in succession in the one season. Because shading of clover is more likely early in the season there is much to be said for fairly lax strip grazing at this time in order to go over a field fairly quickly.

There is another sound reason for adopting this procedure, especially in districts where cold spells of weather are commonplace in the spring. If a pasture has attained some length and is grazed very bare, the onset of unfavourable weather conditions will result in very poor recovery, but if some leaf is left after grazing the pasture keeps growing. In order to counter the ill-effects of fouling by this rather extensive form of grazing, such pastures are closed after two visits by the herd for late-spring silage at a time when clover is better able to compete with grass than it is earlier in the year. Under this system, the electric fence comes into its own from about the beginning of May, when most fields are closed for conservation and when grass growth is rapid.

Invariably the pastures which beat the brunt of fold-grazing during May and June open up to some extent, and by the beginning of July they will carry some rough patches which are largely the result of fouling. Sometimes such fields are closed for a light crop of late hay or silage, or else they are hard grazed by the ewe flock after weaning. The autumn flush, needless to say, is rationed carefully with the electric fence to avoid wastage.

At Cockle Park we were fortunate in having sheep in conjunction with dairying, because they are so useful in combating any opening-up effects in the sward due to the grazing by dairy cows. Generally, if a sward becomes open and there is a loss of clover, often aggravated by a fairly heavy use of nitrogen, it will become a fat lamb pasture for a year. This has a further advantage so far as the lambs are concerned, in that one is certain that they are on clean pasture.

Farmers who have no sheep and do not practice a strict two-sward system can at least use the mowing machine to keep their pastures fresh and in active growth. Wherever intensive stocking is practised the endeavour should be made to alternate grazing and mowing, in so far as this is practicable. However, mowing should also be part of the conservation plan, so that two purposes are served, and this plan should be formulated early in the year. One or more fields which are not grazed at the beginning of the season are harvested as the first silage to be made and, cut at the flag stage, these will produce a good aftermath which will come into the grazing rotation as fields that have been grazed more than once are laid up for conservation and cleaning.

There may have to be deviations from the original plan as the season progresses. More or less grass may be conserved than was originally intended, but the overall aim must be to avoid any waste of grass or of its growth potential so that adequate reserves of winter fodder are created. Anyone who wishes to expand output cannot have too much silage or hay. Once this point is reached, upward adjustments in stocking can be made with confidence because the feeding prospect is secure for the future.

STOCKING POLICY

What is a reasonable concentration of stock on a mainly grassland dairy farm? No firm answer can be made to this question, because of variations in climate, soil, and fertiliser practice. On good deep soil, however, with a satisfactory rainfall distribution (or with irrigation), it is now reasonable to aim at a cow equivalent to 0.4 ha of grassland and forage crop, with not more than one tonne of homegrown or purchased concentrates per cow equivalent, unless yields per cow are appreciably above average. A cow equivalent is defined as any animal that enters the milking herd, or any two animals in the process of rearing, excluding calves under three months, which are still on milk or mainly purchased food.

It is important to keep rearing stock to a minimum, especially on the smaller farm, because they are maintained largely at the expense of milking cows, which make a more remunerative use of food. This reflects the importance of calving heifers for the first time at an early age, certainly not later than $2\frac{1}{2}$ years, and of maintaining a high proportion of mature cows in the herd. These are points of general farm management rather than grassland management, but nevertheless they are very important in securing maximum profit from grassland.

RATIONING OF CONCENTRATES

Returning again to the question of levels of concentrate feeding, where there is a main reliance on pasture we prefer not to think in terms of getting maintenance and x litres from grass or bulk foods and feeding concentrates for y litres. This is partly because the practice makes a distinction between nutrients from different sources which a cow's metabolism does not necessarily recognise, and partly because it does not emphasise sufficiently the economic significance of level of concentrate feeding in relation to yield.

For example, if one is told that a Friesian herd is averaging 18 litres of milk daily with the use of 0.35 kg of concentrates per litre during the winter months, there is the knowledge that reasonable standards are being achieved. If, however, the performance falls appreciably below this standard, one should start to examine the management factors responsible for the decline. Possibly there are too many stale cows as a result of a bad calving pattern; yields may be suffering from poor milking methods; or, most important in this context, there may be insufficient quality in the hay and silage. A simple statement of average yield and concentrate use for given periods becomes very valuable in management when one has records over a succession of years. It gives a basis for critical assessment, not only of progress or otherwise, but of management factors influencing the situation.

There is another argument against the maintenance-plus approach to the feeding of dairy cows. One may send a sample of silage away for analysis and get a report back indicating a good enough feed for maintenance and the first 10 litres. When a cow starts to approximate to this level of production and her concentrate ration is suddenly cut off there will be a sharp fall in her production. This will be due primarily to a change in her management. It should not be interpreted that the silage is necessarily below the value stated and that there must be a general raising of concentrate feeding to all cows.

Except where cows have reached the point of drying off, it is almost better where one is relying largely on bulk foods in the winter to think in terms of feeding say 0.2 kg of concentrates/litre to the cow giving 10 litres, 0.25 kg for the cow giving 15 litres, and so on up to a maximum of 0.45 kg per litre for the cow giving 35 litres or more. This is a different approach from the conventional one, but it is rather more appropriate to the farm where a conscientious effort is being made to provide good hay and silage, and when there is a high level of herd management.

It may be possible if the bulks are of exceptional quality to provide lower rates of concentrate feeding or it may be necessary to raise them a little if the bulks are not up to standard. In fact, a farmer is advised to experiment a little on his own behalf when he starts a clamp of silage to see what level of concentrate feeding he should be following, using cows that have settled down in their lactations.

Differential feeding of concentrates in this rather precise manner may be a little complicated for the large herd which is handled by a small labour force. Here a farmer may elect to feed a flat rate per litre for all cows giving 7 litres of milk or more during the winter months, the rate per litre being varied each year according to the quality of silage and hay available. The quality of the herd and the herd management would also influence this decision, inasmuch as a man with a management herd potential of 4,500 litres per cow would be justified in feeding a much higher rate than the man with a 3,500 litre level of herd and management.

Such a flat rate of feeding does not cater for the exceptional cow that is unable to manage a large intake of bulk foods. But this sort of farmer is less interested in the individual performance than he is in the average performance of his herd, although he must always be looking critically at his results in relation to rising costs.

The greatest economies in concentrate use can be made in the grazing part of the dairying year, especially from May on to the start of autumn calving. It is not suggested that there should be a sudden cutting off of concentrates when cows go out to early grass, as this may be positively dangerous because of grass tetany or staggers. Wherever there is a risk of this, it is advisable to continue feeding some concentrates fortified with calcined magnesite to all cows giving 15 litres or more daily, say at the rate of 0.2 kg per litre until the feed hardens in early May. This need be no more than a grain or dried sugar beet supplement, because excepting the really high yielder, any cow getting a belly full of spring grass twice a day will be receiving all the protein she needs. Economies can be made not only by feeding less concentrates but by substituting cheap starchy foods for expensive protein when there is ample grass protein in the diet.

A question left unanswered is what one does with a 35 litre spring calver on grass in May. Our answer is to feed cereals after 25 litres at some 0.5 kg per litre.

In our experience, two things may happen if this cow is not fed. Firstly the cow will lose flesh rapidly and consequently will be extremely difficult to get in calf while she is losing weight, and secondly she will rapidly come down from her peak yield and often settle at some 18–20 litres, in fact, lower than a herd mate who may have peaked at only 25 litres per day.

We have, however, a high regard for the adequacy of spring and summer grass for cows that have been in milk for five months or more, but the value of autumn grass can be over-estimated, especially for freshly-calved cows. These may be milking well at the outset of lactation but largely because they are milking off condition. Once a cow loses condition and falls sharply in her production at this time of the year, no amount of subsequent good feeding will restore the level of yield. It is better to think of autumn grass for this category of cow as one thinks of good silage later in the year, and feed accordingly. It is a different matter, of course, with a late-winter calver which is nearing the end of her lactation, for it is unlikely that worthwhile response will be obtained from concentrates if the autumn grass is of reasonable quality.

CHAPTER XX

SUMMER OR WINTER MILK PRODUCTION?

Contrasts Between Britain and New Zealand

THE majority of British dairy farmers aim to calve their cows in the autumn to take advantage of the higher prices that have prevailed for milk during the winter months. This pricing policy has been dictated by the need to ensure that there are adequate supplies for the lucrative liquid milk market as opposed to the less remunerative outlets such as cheese and butter manufacture which have had to compete with cheap sources of supply, either the highly subsidised countries of Western Europe or such a country as New Zealand where the natural and economic conditions make it possible to produce manufacturing milk very cheaply. The New Zealand dairy farmer would count himself very fortunate if he were getting as much as half the price that his British counterpart has been receiving in recent years.

Calving on a typical New Zealand factory-supply farm starts just prior to the onset of active pasture growth, for experience has shown that this pattern of calving results in the cheapest and most efficient system of production under New Zealand conditions. The great majority of the dairy farms are family concerns and it is common for no more than two people to look after herds of 100–150 cows and do all the other work associated with the running of a farm. It is understandable therefore that farmers aim to develop a system that not only minimises labour effort, but also gives them a break of at least two months from the double daily chore of milking cows.

They secure this objective by seasonal production. Through calving in the early spring they ensure that the peak nutritional demands of the herd in the four-month period after calving coincide with a period when grass is at its best, nutritionally. As pasture starts to fail both qualitatively and quantitatively in the autumn, the herd is approaching the drying-off stage. There is a relatively long dry period of at least sixty days when the herd will be maintained on hay or silage which is essentially a by-product of the intensive grazing management practised during the main growing season. There is nothing like the same emphasis placed on the quality of conserved grass that one finds on the better grassland dairy farms in Britain which have to cater for the winter needs of herds that are at a peak potential for milk production.

There is a big emphasis on stocking intensity and most of the better New Zealand dairy farms will carry a cow equivalent or better to 0.4 ha, without any purchased food. Much more weight is placed on yield per ha than on yield per cow. In other words, in order to maximise utilisation of pasture, farmers accept a reduction in yield per cow as a consequence of high stocking rates, provided their overall objective of high production per unit area is achieved. Even with this, because of good cows and good stockmanship, it is not uncommon to find herds of Friesian and Friesian cross cows which will average 4,000 litres of milk or better, testing over 4 per cent of butterfat. Naturally yields per cow are much lower in Jersey herds, but yields per ha are only slightly less than those achieved by Friesian herds because of the higher stocking intensities that are possible with the smaller breed. There are many dairy farms in New Zealand that produce more than 9,000 litres of milk per forage ha, which is almost entirely grassland, and rear all replacement stock into the bargain.

The New Zealand approach to pastoral dairy farming, with its emphasis on utilisation *in situ* and a very different calving pattern, is in marked contrast to that in Britain where the majority of cows are on diets that are strongly fortified by concentrates in the winter months when they are in full milk. When they go out to pasture they are mainly stale cows and though there is usually a lift in daily yields it is only rarely that autumn-calving cows will average appreciably more than 25 litres per day when they are out at grass. The calving peak is not so distinct as it is in New Zealand. Though most farmers calve their replacement heifers in the early autumn, inevitably there is a slip back in calving dates, particularly where cows are tied by the neck in byres. Heat periods are of a relatively short duration in the winter months, especially where nutrition is inadequate, and it is not surprising that herdsmen often fail to observe that cows are in season. The inevitable consequence is that most of the so-called autumn-calving herds have a calving spread that extends from the late summer to the late winter and as a result dairymen are milking cows every day of the year.

In contrast, the typical New Zealand factory supply herd has a calving spread of no more than 6–7 weeks. This is not achieved by heavy culling because annual herd wastage from all causes—low production, failure to breed, disease and accidents—is below 20 per cent. The simple fact is that it is easier to get cows in calf in the early summer than it is at any other time of the year and there is seldom any difficulty in determining which cows are in season when they are out at pasture and they are being mounted by other cows.

SUMMER MILK PRODUCTION AT COCKLE PARK

It is not surprising that the senior author, with his New

Zealand background and a feeling of certainty that Britain would be in the Common Market before many years were out, turned his thoughts about the mid-nineteen sixties to the possibilities of developing a milk production unit at Cockle Park that was very much on the New Zealand pattern. He argued that if there was going to be Common Market prices for butter and cheese, which are both storable commodities, and also Common Market costs for concentrates, it would be better for Britain that the highest possible proportion of manufacturing milk should be produced cheaply from grass as grazing, rather than from conserved grass that has to be supplemented with expensive concentrates.

Cockle Park, with its cold heavy soils and late springs, is in one of the less favoured parts of Britain for summer milk production but it was felt that if the project succeeded there the point would be made for other regions of Britain notably the West and South West, where the growing season is much longer and it is possible for herds to have reasonably good grazing for up to nine months in the year.

The unit was established in 1969, mainly with a herd of two-year-old Jersey heifers. By 1972 a more normal age balance was achieved in the herd of 48 cows which was run on a block of 15 ha of which 3 ha was in barley undersown with RvP ryegrass for the 1973 grazing season. The remaining 12 ha were all in ryegrass and these supplied the grazing needs of the herd as well as 180 tonnes of wilted silage for winter feeding. The 3 ha of barley were not sufficient for the total cereal needs for the herd and so a small quantity of barley had to be "bought in" from the main farm. The only other purchased food was high magnesium pencils which were fed as a precautionary measure against hypomagnesaemia until about the end of May. The barley straw, with an appreciable amount of Italian ryegrass in it, is a valuable source of additional nutriment for the cows just after they are dried off.

In 1972, the last year of this project, a stocking rate of a cow to 0.25 ha of grass was maintained throughout the season and the concentrates usage per litre (almost entirely barley) was 0.77 kg.

The yield per grassland ha was 11,900 litres as compared with 11,700 litres in 1971. Comparatively, 1972 was a much poorer grass year with a late cold spring, a lot of early summer poaching, and drought from the beginning of August to November. As a consequence of a fore-shortened lactation averaging 270 days milk yielded at 2,950 litres per cow, was appreciably lower than one can expect even with such stocking rates in a normal season. The target figure was in fact 3,200 litres per cow which was a reasonable expectation for an age-balanced herd of Jerseys, that had been subjected to selection, under these conditions.

Nevertheless the financial results in 1972 were impressive with a gross margin of £524 per grass ha and £405 per forage ha. This last figure was obtained by adding to the total area of the unit, the acreage equivalent of the "purchased" barley. These margins were nearly double the normal returns at that time.

FRONDEG FARM IN CARDIGANSHIRE

Not to be outdone by Cockle Park, the junior author, when the Welsh Agricultural College took possession in 1970 of Frondeg Farm, decided to make summer milk production the main enterprise. He chose Friesians and he had more than a personal attachment to the breed to justify his choice because, the need in Britain is for dual-purpose cows even if the quality of their milk for manufacturing purposes is not all that is desired. By 1976 the farm of 56 hectares carried 105 milking cows and 66 head of replacement stock and it also provided winter grazing for 250 breeding ewes. The herd now averages 6,069 litres of milk sold.

It is early days for this farm, which is still very much in the process of development. The herd is still a young one and there has not yet been an opportunity to select cows and many in the herd are in the category of make-weights until such time as they can be replaced by stock of greater potential. The level of fertility of the farm was initially very low and in our experience, no matter how liberal one is with lime, phosphate, potash and nitrogen at the beginning, it takes several years before pasture land is brought into full heart. Both Stapledon and Levy, the high priests of grassland farming in Britain and New Zealand respectively, stressed the importance of "stock fertility", which implies the build-up of the biological component of soils as a result of producing more grass and carrying more stock with an enhanced return of excrements that add to biological activity in the soil. This has never been properly defined but yet it is a very real phenomenon. One has only to see New Zealand grassland farms that have doubled and trebled in carrying capacity over the years to appreciate the dynamics of grassland improvement which cannot be accounted for by the fertility status of soils when expressed purely in terms of conventional chemical analysis of plant nutrients.

WOORE HALL, SHROPSHIRE

Even at its present level of production, the figures for Frondeg are good and give support to our thesis that it is the western part of Britain that holds the most promise for summer milk production. But this is not a normal commercial farm, nor is Cockle Park, in the sense that those who are responsible for management do not depend on the farm's profits for their living, and so we turn to a genuine commercial, enterprise, that of Charles Platt of Woore Hall, near Whitchurch, Shropshire, who several years ago—well in advance of contemporary thinking, let alone practice—decided to go into summer milk production.

By any standards the level of this farm's performance is outstanding. Its 75 ha carry 162 cows which in 1976 averaged 4,889 litres of milk sold with an average of 0.46 kg of concentrate fed per litre. Here we see in a commercial perspective the New Zealand system of grass farming with modifications to meet the differences in climatic and economic conditions that exist between the two countries, in particular the relatively heavy use of nitrogen (380 kg of N per ha) because a complete reliance of clover nitrogen under these conditions would not be able to sustain the level of stocking that this farm now supports.

But this performance is not obtained by nitrogen alone. There is a herd of first-class cows and the genetic quality of the herd is being enlarged by the use of nominated sires in a well-founded herd improvement plan. We stress again that good cows and good cowmanship are integral parts of good grassland dairy farming, which are just as important as good pastures and efficient grazing and conservation practices.

Charles Platt set stocks his pastures, rather than paddock graze them, and this is right for he is one of those people who have both the dedication and the expertise to make the most of grass in this way rather than by the rule-of-thumb paddock approach we recommend for people with a lower level of competence in matching the needs of their herd to the level of grass production by adjusting the "set stocking" area.

He makes silage on the Dorset-wedge system and this suits his management plan very well for it means that the last made silage, which even with one's best effort is usually nutritionally inferior to early season silage, is on offer to the herd under a self-feeding regime when the cows are dry, while the better quality silage is on offer during the final steaming-up stage and in early lactation before sufficient grass is available to sustain the level of yields. This does not imply, however, that one should deliberately sacrifice quality for quantity in the silage-making programme in order to achieve this end. Charles Platt admits that he made this mistake in the early years of his venture, but he found that it was better to make the effort to produce as good silage as he could for his cows during their dry period so that they calved down in the sort of condition that promoted a good early flush of milk that could be maintained by good silage and limited concentrates until such time as adequate pasture was available. Such a precautionary policy will pay dividends if there is a late spring because once a cow starts to drop in yield as a consequence of poor nutrition it is very difficult, if not impossible, even in early lactation, to bring her back to her potential yield level.

Because he is a commercial farmer, whose private business affairs must be respected, we have not asked financial details of his enterprise but we can say that his approach, with his expertise as a manager, gives him more confidence in the future of his farming, with entry into the Common Market, than the great majority of British dairy farmers who stick to the conventional approach of winter milk, combined with a liberal use of concentrates can possibly have, now that concentrate prices have risen so steeply. He has achieved a truly low-cost system of milk production that can be a model for other progressive farmers to follow. Admittedly he has a demanding time in the late winter at the peak of calving and again in the spring when the risks of hypomagnesaemia are at their greatest, but once the herd is safely out to grass both the labour burden and the anxieties are eased. Then there is the solace of easy days in December and January when the cows are dry and self-fed silage is sufficient for their needs.

PROBLEMS OF SUMMER MILK PRODUCTION

It must not be assumed from the foregoing that a visit to Woore Hall, Frondeg or Cockle Park will provide all the answers to someone wishing to follow the trail that these farms are blazing because there are still a number of unresolved problems and unanswered questions. For instance, there is a suspicion that the feeding of magnesite to combat the dangers of hypomagnesaemia in the spring impairs conception, and it is possible that here is a veterinary problem yet to be resolved.

Again we can be by no means certain about the levels or the duration of concentrate feeding that should be adopted to get the most economic response. The weight of experimental evidence relating to cows fed concentrates when they are on good grass indicates that for the duration of the trials, the response to concentrate feeding is insufficient to justify its use. Some years ago Dr. Don McClusky, while he was at Newcastle, reviewed both Dutch and British literature and came to the conclusion after examining the results of more than thirty trials that it took a similar number of kgs of concentrates to produce an extra litre of milk as compared with cows that were fully fed on grass alone.

But these results referred to the duration of the experiment and average cows. They took no account of freshly calved cows which, when fully fed, were capable of producing 35–40 litres per day, the residual effects of good feeding early in lactation on yields later in lactation, nor the ups and downs of pasture production under farm conditions. It has been well established that peak yield, provided it is accompanied by adequate nutrition subsequent to this peak, has a marked influence on the total yield for a lactation. Expressed in another way, an identical twin with a low peak yield as a consequence of inadequate feeding prior to and subsequent to calving will not in later stages of lactation match the level of production of its twin mate that has had a higher level of nutrition at these critical times, even when they are subsequently given the same level of nutrition. Just as the child is father to the man, so then does feeding of a cow before and after calving have a profound effect on total performance during the subsequent lactation.

It is for this reason that it has been decided at Frondeg, to investigate the value of concentrates for high-yielding spring-calving cows that depend principally on pasture as their source of nutrients. We have no pre-conceived ideas about the possible economic results of this exercise. It may well be that we will have to accept a lower level of yield from spring-calving cows if the additional yield of milk that is obtained from feeding concentrates does not cover the cost of this additional feeding.

It is possible, too, that there is an important individual cow reaction to a complete reliance on pastures inasmuch as some cows may be superior to others in making the best of the grass that is offered them. There was more than a hint of this in the observations Hancock made at Ruakura in New Zealand on the grazing habits of identical twins. High-yielding twin pairs characteristically spent a longer time grazing than low-yielding pairs and were more selective in the grass they ate. In that country, where there has been selection for high yields with pasture feeding over many generations, it is not uncommon for Friesian cows to average more than 18 litres daily over their whole lactation without any assistance from concentrates. If the price of the latter are high in relation to milk prices, we may be forced into very low level concentrate feeding, and if this is so, then concentrates should be fed early in lactation when a cow is at her peak of efficiency in converting nutrients into milk, rather than after her peak when additional feeding tends to put fat on her back rather than milk in the bucket.

We cannot be categoric about the optimal time for calving cows. We suggest about six weeks in advance of active pasture growth as being a suitable time, but in the more favoured pastoral areas such as Pembrokeshire or County Cork it may be preferable to reduce this interval to a month and thereby effect economies in the conservation programme along the lines adopted by New Zealand dairy farmers. Calving in April, in our experience, is too late because we find that though these late calving cows have a good peak yield, they have a fore-shortened lactation.

WINTER MILK PRODUCTION

We are not because of our special interest in this topic advocating a wholesale move to summer milk production, since, for obvious reasons, winter milk production will continue to be an important feature of British dairying for many years to come. What we are suggesting is that any expansion of the industry as a consequence of a greater demand for manufacturing milk should mainly be in the summer production category, especially in those parts of the country that are well suited to this approach. The great majority of British dairy farmers, however, will still be concerned with the problems of producing winter milk as efficiently as possible in the face of rising costs, especially those for land, labour, and purchased feeding stuffs.

We can do no better, in establishing the basic principles we believe to be important in developing a fully competitive system of winter milk production under Common Market conditions, than describe the example of a quite remarkable enterprise, Watson Hill, near Egremont on the West Cumberland coast, which is farmed by Edwin Bushby, a past President of the British Grassland Society. It is not enough to know the farm as it is. To put the whole achievement in perspective it is important to know something of the farm as it was some twenty years ago when it carried about forty Northern Dairy Shorthorn cows with a very average yield and it had pastures no better than those of neighbouring farms.

In 1976 this 50 ha farm carried 126 milking cows which averaged just over 6,400 litres of milk sold, with an average concentrate use of 0.21 kg/litre. These concentrates consisted mainly of cereal nuts because the invariable quality of the grass and the silage is such that starch and not protein is the required supplement. The farm breeds its own replacements, the progeny of nominated sires, but these are reared on another farm. The gross margin is at a very high level of £906 per hectare, and because labour utilisation is efficient the net margin is such that there would have to be a substantial erosion of milk prices before Watson Hill would feel the draught. The efficiency of this small unit has meant that it has generated sufficient capital in the past three years to allow Mr. Bushby and his sons to acquire adjoining farms.

The key factors in the success of this farm are, first of all, a good herd of Friesian cows which are subjected to first-class management. One has only to walk through the cows to realise that they are members of a happy herd and here again we stress the point that good stockmanship is basic to the task of efficient conversion of feed into milk. Secondly, there is a carefully determined programme of pasture production which utilises in full current technology in this field. Thirdly, there is efficient utilisation by controlled grazing which is integrated with conservation to provide a high quality silage for winter feeding of the autumn-calving herd.

Strip grazing is practised but the fields are fairly small, so in effect the system exploits the best of two worlds because the stock are given the grass they need without the detriment of excessive back grazing of recovery growth. In the process, management retains the flexibility which is so important in matching the needs of the herd to fluctuations in the availability of pasture.

The conservation programme is based on silage which is cut prior to heading to provide a digestible product with a high protein content. In order to maximise intake the material is wilted, usually for 24 hours, before ensiling. Year after year Watson Hill is remarkably successful in producing a silage that the cows really enjoy and as a result supplementation is kept to a reasonably low level.

There is no parsimony in the use of fertilisers. Some fields receive as much as 500 kg of N per ha, but the average for the farm is much less because there is an opportunist use of nitrogen to ensure that the sequence of growth is adequate for both grazing and conservation. This is one of the striking features when one visits the farm in the summer, for always there seems to be grass in plenty, despite the heavy stocking rate. There is little manifestation of waste in the form of neglected patches because mowing is integrated with grazing not only for conservation purposes, but also to preserve the palatability of pasture.

This is a farmer's observation, without scientific evidence to support the view, but Edwin Bushby believes that the application of potash from May onwards adds greatly to the palatability of grass. He is of course wise to delay these dressings to the summer because potash and nitrogen in the early spring at the rates used on Watson Hill would be a sure recipe for hypomagnesaemia.

About 20 per cent of the farm is in Italian ryegrass, which is normally retained for 18 months. Both spring and autumn seedings are practised, with a spring-sown block being broken in its second summer to establish an autumn-sown crop which makes an early contribution to grazing in the following spring. The spring reseed is of course especially valuable for the later calving cows, to give a boost to milk yields during the "summer gap" when older pastures are at a trough both in production and palatability.

The remainder of the farm is in long-term leys which are only broken for reseeding when this is deemed to be the wisest course of action. Timothy-meadow fescue pastures have been favoured and it is quite remarkable, despite the very high level of nitrogen usage, that white clover becomes a dominant component of these swards during the summer months, even to the point where it can be considered to be a weed because it does not respond to nitrogen applications. One reason for breaking a pasture is that it has become clover dominant. Partly this persistence of clover is attributable to the management system which never allows grass to become mature and shade out the clover component of the sward but there is more to the story than this. Neither timothy nor meadow fescue is an aggresive producer of tillers in the way that ryegrass is. Because clover is a light demanding plant it has a compatability with timothy and meadow fescue with intensive utilisation which minimises shading effects, but the same regime will have a very different effect with a clover-ryegrass association.

Some may ask whether Watson Hill would be an even more productive farm with perennial ryegrass instead of timothy and meadow fescue. The answer possibly is 'yes' but the fact is that Edwin Bushby has found that his combination is a very satisfactory one under his conditions of management and generally it is not wise to move from a proven system or practice until there is more than marginal evidence that a change is justified. Nobody could ever accuse Edwin Bushby of being complacent because he is one of those people for whom perfection is always around the corner to be achieved by yet more effort. We know very well when this book is revised in a few years' time that Watson Hill will have an even more impressive story to tell.

THE FARMER'S CONTRIBUTION

This brings us to the final point we wish to make in this

chapter—the enormous contribution that practising farmers have made to the technology of grassland farming, whether it be in milk, beef or fat lamb production. Most scientists with an interest in grassland, work in fairly narrow fields such as plant breeding, plant nutrition, ecology, or ruminant nutrition and they are seldom able to see the picture as a whole where soils and plants and animals form the complex which the farmer must manage to the best of his ability to make a worthwhile living from his land. One criticism of grassland research in Britain, certainly up until comparatively recent times, is that there have been too few scientists who have appreciated the fact that grass, apart from any amenity value it may have, only becomes meaningful when it is turned into milk or meat or wool.

Fortunately for Britain this gap in development, as opposed to research, has been filled by progressive farmers and this is one of the great strengths of British agriculture with its elite corps of pioneers and innovators. One has only to work in a static system of farming, as the senior author has been in Spain to realise the value of the contributions leading British farmers have made to their own industry.

Go back forty years and the name Hosier was a living reality because he brought an example and a new dimension of thought to our industry. He was followed by Rex Paterson, the master dairy farmer of our times. Then there were men like Sandy McGuckian in Northern Ireland and Captain Keith and Maitland Mackie in Scotland who left their imprint on farming practice.

In the last edition of this book we told the story of Oliver Barraclough, on poor land at the 300 metre contour of the hungry Pennines near Bradford, who took his dairy farm from mediocrity to a level of production that even the best lowland farmers must envy. We also described the combined efforts of the late Edward Moffitt and his son John, better known for their remarkable Hunday herd of Friesians, for the development of West Newham in Northumberland. These men were selected not because they are unique but because they exemplified the best in

British farming—a capacity to use new knowledge to the betterment of farming systems.

Nowhere is this ability to relate knowledge and effective execution more important than it is in grassland husbandry. The proprietor of a broiler unit or of an egg-laying factory works to a recipe, and this is almost true of the arable farmer who is concerned only with cereals and a couple of break crops like sugar beet, vining peas or potatoes. But the grassland farmer has to work with two living entities, his pastures and his animals, to achieve an economic result. All of us concerned with the future of British farming cannot but be indebted to Charles Platt, Edwin Bushby, Fenwick Jackson, and men of their ilk for the contributions they are making to the quality of our grassland farming.

CHAPTER XXI

THE FUTURE OF GRASS FARMING

IT is not easy to predict the shape of things to come in British farming but of one thing we may be reasonably certain—the next thirty years will see even greater changes than we have experienced over the past thirty. Apart from technological advances, of which there will be many, there will be substantial structural changes resulting in greater specialisation and larger scale production. Agriculture will not be unique in this respect for the same trends are occurring in other industries. Higher labour productivity, on which higher material standards of living largely depend, will require greater capital inputs in the shape of machinery and fixed equipment and this will necessitate an increase in the size of production units in order to effect economies of scale.

Another factor contributing to greater specialisation will be an increased need for management expertise of the highest quality. There are few men blessed with the competence to be authoritative and efficient in many branches of present-day farming and so there is a necessity for a greater concentration of effort in order to achieve the highest possible rate of technological advances. This is specially true of grassland farming with its complex of pasture and animal management.

There are two over-riding influences which will have a profound effect on our farming. The first is the absolute rate of increase of the human population which will eventually reduce the amount of food that is now directed to this country. The second is our entry into the enlarged European Economic Community. Generally speaking, British agriculture has been able to face this event from a strong position because of structural advantages as compared with most of the Continental countries. The large-scale arable farmer in eastern, central and southern districts has nothing to fear because he is in a highly competitive position and is enjoying higher prices for cereals than those received before entry. Unquestionably too, any farmer operating on a reasonably large scale in regions well suited to pasture production will be in a very strong position if he utilises his grassland efficiently. The immediate effect of high cereal prices will be an increase in the area devoted to cereals, at the expense of grass, but this will not be a bad thing because, as happened during the Second World War, there will be greater incentives to make the most of the remaining grassland.

DAIRY FARMING

In these developments the dairy industry will have a major part to play, especially on lowland farms in the higher rainfall districts, because we will need an expanded national dairy herd not only to make good the shortfall in butter and cheese, formerly obtained from Australia and New Zealand, but also to provide, as a by-product, many of the additional beef stores that are required to meet the rising demand for beef within the Community that cannot be satisfied by imports. This expansion in dairying must be achieved by a greater relative use of grass nutrients for we have said goodbye to the favourable relationship between milk prices and the cost of concentrates that existed under the cosy arrangements of our old guaranteed price system. We have moved into a situation where we have Common Market prices for milk, which though higher than those obtaining before entry, are accompanied by Common Market prices for cereals which are very substantially higher.

Our top grassland dairy farmers certainly have no need to be dismayed about their future because they are already in a very strong position. One is thinking here of the men with 100 cows on 50 ha of grassland, who are using 350 units of N per ha and are using less than 0.25 kg of mainly starchy concentrates per litre of milk without appreciable detriment to yield. We are thinking of men who seek to emulate the performance of pathfinders like Edwin Bushby, whose farming is described in Chapter XX.

We put great stress on the use of starchy as opposed to protein balanced concentrates in the supplementary feeding of dairy cows that are largely maintained on pasture and its products which, under good management, require minimal protein supplementation. The blind recipe of a maintenance and 5 litres of milk from silage and 0.5 kg of balanced concentrates for each succeeding litre can lead to a serious waste of protein which, in the form of oil seed residues, is becoming prohibitively expensive. Here, indeed, is an area where dairy farming in Britain has been prodigally wasteful and it cannot afford this any longer.

Even the 30 ha family dairy farm can hold its own economically if there is a concentration of effort on growing grass and turning it into milk for by Continental standards this is a relatively large farm and it will take many years for the Continent to effect the necessary structural changes to make farming more efficient. There were ten million people less on farms of the Six at the end of the sixties than there were when the Common Market was introduced, but this process still has a long way to go before the average Continental livestock farm will match its British counterpart.

Meanwhile our leaders in grassland farming practice will have taken further steps to improve their position in the league because we are very far from a plateau in grassland farming practice. One seldom encounters a farmer who is using more than 300 kg of elemental nitrogen per ha and yet it has been shown experimentally that we can use up to 600 kg of N per ha without any sign of diminishing returns in terms of dry matter production, always providing there is adequate moisture, which is not normally a limitating factor in western districts which offer the best prospects for really intensive grassland farming.

CONSERVATION

High yields of grass, however, have no meaning unless they are turned into correspondingly high yields of milk and meat and here we come up against the problems of efficient utilisation. Grazing practices have improved enormously but there must be a matching improvement in grass conservation. The continuing preoccupation with hay, at the expense of silage, largely because it is easier to feed in outmoded layouts such as byre accommodation, not only results, in this uncertain climate of ours, in an end-product of very dubious quality but also it gives poorer aftermath grazing than that of a silage regime. Even a lot of the silage that is made lacks a real concentrate sparing function. We must have a form of conserved grass that is capable of supplying four-fifths of the nutrient requirements of the average cow in milk.

This can be obtained from really high quality silage, with a high dry-matter content or from dried grass. Dried grass failed in the past because of the physical inefficiency of the drying process and poor field organisation which resulted not only in high processing costs but also a low level of output leading to over-mature herbage of insufficient quality to justify the high costs of drying, which are such that drying is not a proposition unless the final product has the characteristics of a concentrate.

There have been considerable advances in efficiency both of grass driers and field machinery for collecting grass but these are expensive items which entail further capital expense in the form of processing equipment for converting the dried grass into cubes, pellets or wafers to reduce storage space and to facilitate rationing. When one adds up the costs of all these items and the necessary buildings to house them, a very large sum of money is involved, certainly very much more than a farm with 80 ha of grass and 200-cow equivalents could possibly justify.

We see a future for grass drying but not as an integral part of the grazing and conservation programme of the typical livestock farm with an annual requirement of 100–200 tonnes of dried grass. It seems that an output of at least 2,500 tonnes of dried grass per annum is the minimum in order to obtain the economies of scale that are necessary to carry the overheads of the capital investment. This is the output of at least 200 ha of well-managed grassland. It may not be necessary for the grass-drying farm to have as much grass as this if mutually satisfactory arrangements can be made with adjacent farmers to supply additional grass to keep the plant in full operation over a drying season that extends from early April to mid-October. With a plant of this capacity, 80 ha of grassland will supply all the material required from the beginning of May till the end of June. It is the other months that require the additional land.

There could be a future for co-operative grass drying, but only if the participating members were prepared to accept the discipline required of suppliers of such an operation as the deep freezing of vining peas where very specific instructions are given to farmers. In order to keep a grass-drying plant going over a long season, there must be late summer reseeds of Italian ryegrass to supply herbage for drying in early April, followed by S 24 ryegrass in early May and S 23, or a similar variety, in late May. Cocksfoot, with its capacity to give a relatively stem-free aftermath in June and July, also has a place and so have lucerne and grass combinations to fill the summer gap. The direct spring reseed also has a value at this time. A grass-drying co-operative will never get off the ground if all members have the same sort of grass coming to the harvest stage at the same time.

Unquestionably this is a complex and capital demanding system of grass conservation but there can be no doubts about the value of good quality dried grass for milk production when fed in combination with barley.

UTILISATION *IN SITU*

However, there is another approach to the utilisation of grass in Britain which deserves more consideration and that is greater utilisation *in situ*. The New Zealanders are masters of this approach with their highly seasonal production of dairy produce and meat from pasture. Essentially their system is one of obtaining the best possible coincidence between the nutritive requirements of the herd or flock with the seasonal levels of pasture production. In recent years this concept has had an even greater emphasis in dairy farming practice. In the fifties a yield of 400 kg of butterfat per ha was considered to be good and it was achieved by a reasonably high level of stocking and by attempting to maximise yield per cow by extending the lactation period and by a fairly elaborate grass conservation programme. This involved calving the herd about six weeks in advance of active pasture growth and the provision of hay, silage and autumn-saved pasture, with this last item making an important contribution just after calving.

Today there are dairy farms with outputs of 500–550 kg of butterfat per ha. Calving coincides with the onset of the spring flush of pasture, stocking intensities have been greatly increased, and cows are dried off once pasture growth begins to fail. Conservation is kept down to the amount necessary to maintain the herd over the extended dry period and is largely done as a form of topping to preserve the quality of grazing. The net result is that farmers are working much more closely with Nature because cows at peak yields are directly harvesting grass at its most nutritious stage.

As we pointed out in Chapter XX when dealing with summer

milk production, there is scope for this approach in the more favoured grassland regions of Britain and Eire. Even though they do not have the same length of growing season as the North Island of New Zealand, there is the compensating advantage of cheap nitrogen, which can extend the effective grazing season by six weeks. It is not extravagant to suggest that a farmer in Pembrokeshire or Dorset could, from a ha of well-managed grass, support a February-calving Friesian cow producing over 4,000 litres annually with supplementation at the rate of no more than 0.2 kg barley to the litre, with this being fed in the early stages of lactation.

This is the logical way of production of the additional milk that will be required to replace imports of butter and cheese, especially on family farms where the hard-worked owner wants some respite from milking cows during the shortest days of the year. With developments in long-keeping milk, this approach to gross utilisation could extend beyond manufacturing milk. It may well be with high concentrate prices and high distribution costs that fresh milk, delivered daily, will cater for no more than a luxury market. Possibly this is sympathetic thinking, for both authors were brought up on family dairy farms and they feel for anyone that has to milk cows every day of the year.

FAT LAMB PRODUCTION

Fat lamb production based on early spring lambing, as was pointed out earlier in this book, gives one of the best fits between stock appetite and grass growth, but this branch of farming has fallen behind in the sense that it seems to have been bypassed by the great technological revolution that has transformed every other major farm enterprise in Britain. However, there are now techniques available, as well as the economic incentives as a consequence of higher meat prices, to transform the whole character of fat lamb production. We have in mind such developments as in-wintering of ewes to reduce labour costs and minimise poaching, and the combination of field hygiene and strategic drenching with more efficient drugs to permit of very much higher stocking rates than hitherto were deemed to be safe. The production of 450–500 kg of lamb meat per ha is now a practical possibility.

To achieve this objective it is necessary to have relatively small but highly fecund milky ewes that are mated to rams with a high growth potential, because the essence of success for such an intensive system lies in ensuring that the maximum amount of available nutrients is used to create new meat rather than maintain old flesh in the shape of big ewes that average less than a lamb and a half per year.

The probability is that the type of lamb required will change as a consequence of Continental demand. Lamb meat consumption on the Continent is low not because it is not a preferred food but because it is expensive. The requirement there is for a relatively light well-muscled lamb that is slaughtered in no more than store condition because the Continentals, like most British people, do not savour sheep fat. This class of carcass will suit intensive lamb production for when one is producing 30 lambs to the ha it is not possible to put the finish on them that Ministry certifying officers have mistakenly regarded as being essential for the need of the table.

BEEF PRODUCTION

Now that the price of good steak matches that of salmon cutlets, a new dimension has come into beef production, especially in the production of suckled calves. Nevertheless in Britain, with its limited and expensive land, any great extension of beef production must primarily depend on an expanded dairy industry where milk carries most of the cost of producing a calf. An important need here is a longer herd life so that a greater proportion of cows can be mated to high growth bulls like the Charolais to increase the beefing potential of calves that are being raised for slaughter. This also applies to suckler calf production. However much one may appreciate the quality of beef from a 450 kg Aberdeen Angus cross bullock that is slaughtered at 18 months of age, the fact is that, except for a limited luxury demand, there is no place in Britain for animals that are specially bred for beef if they have this mediocre growth performance. The 18-month bullock must be capable of reaching 600 kg so that its cost as a calf is much more widely spread.

Though barley beef now has little more than a historical interest, we must continue to be grateful to Dr Preston for his innovation because through it we have seen more clearly the issues that are important in securing a higher output of beef from grassland. Among these is a high growth potential and the need to ensure that there is no check in growth, for we simply cannot afford the extended store periods that characterised the three-year old bullocks that used to be fattened on rich Midland pastures.

The 18-months system of beef production based on pasture, silage and a limited input of cereals is the sensible compromise between barley beef and traditional grass feeding of bullocks. Well managed, it can give an output of over 750 kg of liveweight to the forage ha, and this is the standard that farmers must achieve to make beef a proposition on lowland grass.

There is, however, another way of making money out of beef on such farms, with crop by-products such as barley straw and pea haulms, and this is by maintaining a herd of breeding cows, but not on a conventional system of 8–9 months suckling that demands a lot of land. It is a wasteful biological process, once the calf has fully developed its ruminant functions, to 'double process' food into milk and then into meat even though it puts a lot of bloom on the calves, a factor of great importance if one is selling them as weaners.

This is much less important if a farmer is finishing his home-bred calves which is the likely case on a lowland arable farm. Here one can safely practise weaning at 5–6 months and this means that for half the year a breeding cow can be largely maintained on a diet of barley straw supplemented with urea and minerals, with a small amount of cereal as she approaches calving. We have practised this system successfully at Cockle Park using Hereford and Friesian cows that had enough milk to rear two calves. Their formal need of grassland amounted to no more than 0.2 ha per head per annum, and on an arable farm this can be part of the grass break that is so valuable in maintaining heart in tillage land.

UPLAND FARMS

Early weaning also makes sense on upland farms but it is seldom practised. Where climate conditions are difficult and effective grazing is limited to the period May–October, it does not make sense to suckle calves for 8–9 months of the year, especially with early calving that necessitates up to three months of supplemental feeding before effective grazing is available. Surely it is better under these conditions to take a leaf out of the New Zealand dairy farmers' book and work with Nature, calving the herd in April and weaning the calves in early October!

But it will be argued that the hill-farmer gets his best returns from big calves and this is true but it is better to get size from breeding rather than size from age! Our thesis is this. If you place breeding cows under physiological stress, as a consequence of lactation, for only six months of the year which largely coincide with availability of adequate pasture, then you can run bigger cows which can be mated to high growth bulls that will produce 200 kg calves at six months of age. Small hardy cows like the Galloway and the Blue Grey have a justified reputation when they are submitted to stresses, but their calves are slow growers. If you minimise these stresses by early weaning, then something like a Simmental cross cow could be a much better proposition especially if it were mated to a Charolais bull. It will not take the lowland feeder very long to appreciate the virtues of calves with this sort of breeding. It is just a question of courage on the part of some hill farmers to make a change in this direction and the rest will follow.

The emphasis on beef stores from upland farms that has developed since Lord Woolton preached his electoral doctrine of more red meat that brought the Tories back into power after the post-war Labour Government has rather obscured the fact that in the absence of discriminating subsidies sheep and not cattle are the masters of the hill because of their better adaptation to difficult environments. This does not deny the fact that under these conditions sheep and cattle are a better proposition than sheep alone but it is a question of the right ecological balance.

Earlier in this chapter it was stated that sheep had been bypassed in the technological revolution that has transformed British agriculture and in no sector is this more apparent than in upland farming. The prodigal son of a Pennine or Snowdonia farmer who returned home after twenty years of wandering would be back in business as soon as he learned the name of the dog, but if he was from a Cumberland dairy farm he would be entering a completely different world. The truth is that upland sheep farming is a very conservative business. Not only are sheep masters of the hill—it seems that they are also masters of the farm because so often they dictate the style of management.

Upland sheep farming is in a vicious circle. The low level of pasture productivity and the hazards of a prolonged winter necessitate the use of hardy breeds. Hardiness implies a relatively low level of productivity because a hill ewe cannot rear 32 kg twins and then come up smiling with another set of twins at the following lambing and the next after that. Low productivity from the flock means that the farmer is not in a position to make investments for the improvement of his farm that will allow him to carry more productive stock, and always he has the sword of Damocles in the form of a catastrophic winter, hanging over his head, limiting what he can do.

We believe that on many upland farms this vicious circle could be broken by the in-wintering of the flock, not in the expensive layouts that the Land Service of the Ministry seem to favour but in inexpensive installations of corrugated iron on poles to provide partial lean-to cover to an otherwise open yard. This will not only reduce winter stress on the sheep but stress on the farmer as well and permit him to carry more productive ewes. Compare the gross returns of a flock that clips 3 kg wool instead of 2 kg and rears 130 per cent of lambs instead of 85 and one then realises the scope there is for the improvement of grazings to the point where they can carry a flock of this enhanced productivity.

These views, you may say, are just supposition and have no basis in fact—and this is the pity. The Ministry of Agriculture has two upland farms which should be studying just this sort of proposition instead of following in the footsteps of the more progressive upland farmers. The unfortunate fact is that essentially these farms are run by timid committees, and few committees can have the courage to be adventurous or plot uncharted seas. The late George Stapledon, to whose memory this book is dedicated, once said he loved mad farmers because they were capable of showing the shape of things to come. We want some of these on our hills for we desperately need to know more of the potential of our uplands.

We maintain that this country cannot afford to let millions of these hectares have no more than an amenity role for urban dwellers when they take refuge from their battery-cage environments. We believe that the best of our uplands must have the dual function of production and playground. In any case, most towns people finds little appeal in the solitude of a wilderness. They much prefer to see sheep and cows and people.

The pity is that the upland sector of British farming was not in a stronger position at entry into the Common Market and this one can blame on the support system that has been operative because it took the shape of a latter-day Danegeld payment in the form of direct subsidies to keep the wolf from the farmhouse door. In the process it has tended to mummify upland farming. A more positive approach to its problems is long overdue and this basically means a drive to improve the production potential of this land. We can only hope that this will be possible under the arrangements that the Brussels bureaucrats will determine for British farming.

THE MORE DISTANT FUTURE

For the next twenty years at least, our primary concern will be growing more grass and utilising it as fully as we can by developing more productive grazing animals that are given the fullest possible protection from disease hazards and are subjected to the best kinds of management consistent with economic output. Ultimately, however, there will probably be a new use for grass. It could well become the basis of a processing industry that will produce reasonably acceptable human food without the intervention of animals. Even with present knowledge and materials we can grow 15 tonnes of leafy grass dry matter per hectare, which contains the best part of $2\frac{1}{2}$ tonnes of protein. This is the protein yield of a 12 tonne crop of beans and it could represent a great deal of grist for the industrial biochemist's mill that turns out synthetic milk or beef-steak. We leave you with this thought of an exciting new industry producing rather dull foods that will have one great virtue. They will help to save man from the perils that he has created through his own fecundity and the steps he has taken to extend his longevity.

APPENDIX I

SEEDS MIXTURES

Note: The higher level of seed rates should be used when it has not been possible to create the fine seedbeds required for pasture establishment. Higher rates should also be used with broadcasting as opposed to drilling, especially with spring seeding.

All rates are given per ha. On the whole, total rates per ha are lower than those usually given in seedsmen's catalogues, because for the main part only certified seeds are suggested.

1. Spring or autumn direct seeding of Italian (18–24 months' duration):
30–35 kg RvP Italian ryegrass or *Grasslands Manawa* (H 1) or certified Danish Italian ryegrass e.g. *Dasas* EF 486, or *Sabrina*.

2. Undersown Italian for stubble feed, early bite and silage (9 months' duration):
23–29 kg commercial or certified Danish Italian ryegrass, *Dasas* EF 486.

3. One-year mowing ley undersown in a cereal:
14–16 kg commercial or certified Danish Italian ryegrass
7–9 kg tetraploid broad red clover* e.g. *Hungeropoly* or *Tetri*
*If there is a severe risk of clover sickness, substitute 2–2.2 kg alsike clover for the red clover, or include 2–2.5 kg of alsike in addition to the red clover. If eel worm risk, substitute *Sabtoron*.

4. Two-year ley for mowing and grazing based on ryegrass:
8–10 kg RvP Italian ryegrass
8–10 kg S 24 or *Grasslands Ruanui*
1–1.2 kg certified tetraploid red clover e.g. *Hungaropoly*
2 kg S 100, Kersey or *Grasslands Huia*

5. Medium-duration ley (3–4 years) for grazing and cutting based on ryegrass for deep soils:
6–8 kg *Grasslands Manawa* or *Sabrina*
7–9 kg S 24 or *Grasslands Ruanui*
6–7 kg S 101 or Kent indigenous ryegrass
1.6–2 kg S 100 or *Grasslands Huia*

6. Medium-duration (3–5 years) for cutting and grazing on medium to light land:

5–6 kg *Grasslands Manawa* or *Sabrina*
6–8 kg S 143 cocksfoot
5–6 kg S 48 timothy
6–8 kg *Sceempter* pasture meadow fescue
2–3 kg certified S 123 red clover
1.6–2 kg S 100 or *Grasslands Huia*

7. Medium-duration ley (3–5 years) for grazing and cutting on all but very light soils:

6–7 kg *Grasslands Manawa* or *Sabrina*
5–6 kg *Comtessa* meadow fescue
4–5 kg *Sceempter* pasture meadow fescue
6–7 kg S 48 timothy
or
1.6–2 kg S 100 *Grasslands Manawa*

8. Cockle Park mixture 4–6 year for general purposes:

12–16 kg S 24 or *Grasslands Ruanui*
or
12–16 kg Kent indigenous or S 23 perennial ryegrass or *Melle*
with
7–9 kg S 143 cocksfoot
5–6 kg S 48 timothy
2–4 kg certified S 123 red clover
1.6–2 kg S 100 or *Grasslands Huia*

9. Medium-duration (3–4 years) mainly cocksfoot ley for light land, mowing and grazing:

6 kg RvP Italian ryegrass
5–6 kg S 24 or *Grasslands Ruanui*
6–8 kg S 26
6–8 kg S 143 cocksfoot
2–3 kg certified S 123
1.6–2 kg S 100, *Grasslands Huia* or Kersey white clover

10. Long ley (7–8 years) on heavy soils—grazing and mowing:

5–6 kg *Grasslands Manawa* or *Sabrina*
10–12 kg Kent indigenous or S 23 ryegrass or *Melle* pasture ryegrass
6–7 kg S 48 timothy
1.6–2 kg S 100
0.6 kg Kent wild white clover or S 184

11. Permanent pasture mixture for good land, mainly for grazing:

3 kg *Grasslands Manawa* or *Sabrina*
7–8 kg S 23 ryegrass
7–8 kg *Melle* pasture perennial ryegrass
3 kg S 48 timothy
1.6–2 kg S 100 or *Grasslands Huia*
0.6 kg Kent wild white clover or S 184

12. Permanent pasture for marginal land reseeding:

6–7 kg certified Danish Italian ryegrass *Dasas* EF 486
12–14 kg S 23 or *Melle* perennial ryegrass
5–7 kg S 143 cocksfoot
5–7 kg S 48 timothy
2–3 kg crested dogstail
1–2 kg S 100 or *Grasslands Huia*
1 kg alsike clover
0.6 kg Kent wild white clover or S 184

13. Lucerne mixture for very light land:

16–20 kg Europe lucerne or *Vertus* lucerne where *verticilium* wilt is a problem.
1 kg S 37 cocksfoot

14. Lucerne mixture for deeper soils:

16–20 kg Europe lucerne or *Vertus* lucerne where *verticilium* wilt is a problem
1 kg S 215 meadow fescue
1 kg S 51 timothy

15. Grazing pasture for pigs:

7 kg chicory
3 kg S 100 or *Grasslands Huia*
7 kg Kent or S 23 perennial ryegrass

16. Herb strip:

3–4 kg chicory
1–1.5 kg sheep parsley
0.15–0.3 kg yarrow
2–3 kg burnet
2–3 kg rib grass
2–3 kg S 100 or *Grasslands Huia*

APPENDIX II

IMPERIAL/METRIC CONVERSION TABLE

WEIGHT

Pounds to Kilograms

lb	*lb or Kilograms*	*Kilograms*
2.20	1	0.45
4.41	2	0.90
6.61	3	1.36
8.82	4	1.81
11.02	5	2.26
13.22	6	2.72
15.43	7	3.17
17.63	8	3.63
19.84	9	4.08
22.04	10	4.53

CAPACITY

Gallons to Litres

Gallons	*Gallons or Litres*	*Litres*
0.22	1	4.54
0.44	2	9.09
0.66	3	13.64
0.88	4	18.18
1.10	5	22.73
1.32	6	27.27
1.54	7	31.82
1.76	8	36.3
1.98	9	40.91
2.20	10	45.46

APPENDIX II

AREA

Square Feet to Square Metres

Square Feet	*Square Feet or Square Metres*	*Square Metres*
10.76	1	0.09
21.53	2	0.18
32.29	3	0.27
43.05	4	0.37
53.82	5	0.46
64.58	6	0.55
75.34	7	0.65
86.11	8	0.74
96.87	9	0.83
107.63	10	0.93

Acres to Hectares

Acres	*Acres or Hectares*	*Hectares*
24.71	10	4.05
49.42	20	8.09
74.13	30	12.14
98.84	40	16.19
123.55	50	20.23
148.26	60	24.28
172.97	70	28.33
197.68	80	32.37
222.40	90	36.42
247.11	100	40.47

LB/ACRE converted to kilograms per ha

lb/ac	*kg/ha*
10 —	11.2
15 —	16.8
20 —	22.4
25 —	28.0
30 —	33.6
40 —	44.8
50 —	56.0
60 —	67.2
70 —	78.4
80 —	89.6
90 —	100.8
100 —	112.0

CWT/ACRE converted to kilograms per ha

cwt/ac	*kg/ha*
1 —	125.5
2 —	251.0
3 —	376.6
4 —	502.1
5 —	627.7
10 —	1255.4
20 —	2510.8
25 —	3138.5
35 —	4393.9
40 —	5021.6
45 —	5649.3
50 —	6277.0

TON/ACRE converted to tonnes per ha

t/ac	*tonne*/ha
3 —	7.5
4 —	10.0
5 —	12.5
10 —	25.1
15 —	37.6
20 —	50.2
25 —	62.7
30 —	75.3

INDEX